When Religious Faith *Collides* with Science

When Religious Faith *Collides* with Science

A Navigational Guide

By Jan M. Long
Foreword by Mailen Kootsey

RESOURCE *Publications* • Eugene, Oregon

WHEN RELIGIOUS FAITH COLLIDES WITH SCIENCE
A Navigational Guide

Copyright © 2016 Jan M. Long. All rights reserved. Except for brief quotations in critical publications or reviews, no part of this book may be reproduced in any manner without prior written permission from the publisher. Write: Permissions, Wipf and Stock Publishers, 199 W. 8th Ave., Suite 3, Eugene, OR 97401.

Resource Publications
An Imprint of Wipf and Stock Publishers
199 W. 8th Ave., Suite 3
Eugene, OR 97401

www.wipfandstock.com

PAPERBACK ISBN: 978-1-4982-8397-7
HARDCOVER ISBN: 978-1-4982-8399-1
EBOOK ISBN: 978-1-4982-8398-4

Manufactured in the U.S.A.   AUGUST 8, 2016

Contents

Forward by Mailen Kootsey | vii
Authors Message to the Reader | ix
Preface | xiii
Introduction | xxv

Foundational Considerations

Chapter 1   On Knowledge and Certitude | 3
Chapter 2   The Madness of Crowds | 15
Chapter 3   Science as a Candle in the Dark | 29
Chapter 4   Coming to Terms with Reality | 49
Chapter 5   Human Frailty: The Hidden Face of Idolatry | 58
Chapter 6   Dilemmas of Faith | 77
Chapter 7   The Spiritual Brain | 95

The Quest for a Theology Connected to the Real World

Chapter 8   The Power of Story | 107
Chapter 9   Paradigms of Meaning | 119
Chapter 10   Thinking Strategically | 140
Chapter 11   Surmounting Tribal Pathology | 154
Chapter 12   Sacred Credibility in the Twenty-first Cenury | 164

Index | 187

Foreword

I met Jan Long at a Saturday morning discussion group on a university campus where about 30 people from different disciplines have for decades met weekly to work through topics in religion and science. As a focus, the group chooses a book recommended by a member for review, and each week the two-hour session begins with a summary of a chapter or two of the book followed by 60-90 minutes of lively discussion. The range of viewpoints among the attendees is very wide, so the discussions are vigorous and hard to wind down at the end of two hours.

I noticed Jan Long first as a regular in the group who followed the threads of discussion closely. About three years ago I spotted a blog article by Jan on science and religion posted on a Web site I check regularly. At the next group meeting I complimented Jan on the article. That led to a series of conversations and a collaboration on a jointly-written series of blog articles on science and religion.

Recently, Jan surprised me again with a draft of a book on science and religion he had been working on for several years. I was eager to read his work and wondered what he would have to say, knowing that his career is in law and management and not science or religion. Differing from the usual philosophical or theological arguments presented from one side or the other, Jan's book turned out to be the story of his personal spiritual and philosophical journey from a conservative Christian home background through a thicket of new scientific ideas that conflicted with the literal Biblical interpretations of his religious roots. Jan's personal journey was of great interest to me because I too grew up in a home with the same conservative Christian denomination. Since my own career was in physical and biomedical science, I could not ignore critical areas of conflict such as the age of the earth and evolution and I too had to make radical changes in beliefs while wanting to maintain a Christian viewpoint.

What I value in Long's book is his description of the array of belief conflicts that he faced, why he found them troubling, and how he has related to them in a thoughtful way. Recognizing that each reader will bring a different background to the search for truth and certainty, he does not attempt to construct a final solution to satisfy everyone. Rather he describes what he calls a "process discussion," describing the pitfalls associated with the different forms of knowledge so that the reader can more safely navigate in constructing a personal philosophy and world view. Long may be a layman in both science and religion, but he has been willing to read extensively and to listen to guidance from friends in both fields. Enough references are provided to give the interested reader a head start in each area.

I think that readers who want to construct a solid personal philosophy and who would not be satisfied by either extreme of atheistic naturalism or conservative Christian literalism will find this book helpful and, as the title indicates, a navigational guide.

Mailen Kootsey
Ph.D. in Physics, Brown University
Academic and Administrative career at
Duke University, Andrews University, and Loma Linda University

Author's Message to the Reader

This book has a very personal origin, given that I was reared in a religious tradition that holds rigidly to the notion that creation of the "heavens and the earth" occurred over a 6-day period roughly 6,000 years ago. At the same time, I am mindful that various aspects of this parochial worldview extends well outside of my specific religious tradition as can be conjectured from polling data. In the Preface I will develop some of this in a little more detail, but for now I am primarily interested in providing some high-level context.

Given that science has come up with very different conclusions about the age and evolution of the cosmos, the question then becomes, "How should an individual go about concluding on such matters?"

This book attempts to answer this by providing readers with a *process* discussion about how knowledge is acquired. The fundamental objective will be to push back against some of the less defensible approaches, doing so by offering up a systematic process that has demonstrable reliability. It is, in fact, this reliability that can afford relative confidence of being on the right track.

The main premise of the manuscript is that there are rules that can assist in navigating conflicts between the sacred stories that deliver meaning and the sense data that keeps humans grounded to the real world. As such, the book sets forth a discussion for those who have a genuine desire to live as connected to reality as possible, providing the intellectual tools needed to navigate conflicts that can emerge between science and religion, recognizing that some parts of every individual's inherited worldview challenges such connections. The proposal I put forth is that there are good reasons for people of faith to maintain respect for sense-based quests that are engaged in systematic inquiry, if for no other reason than the fact that history clearly provides guidance on how the zone of overlap should be handled. This manuscript is largely a search for those principles that can guide a person in the exciting life-long process of getting acquainted with the world in all its incomprehensible permutations.

In order to develop this conversation, it has been necessary to consider the major components that go into human thinking—things like the sources of ideas, structure, content, limits, and vulnerabilities intrinsic to all ideas. It has therefore been necessary to reflect on differing types of knowledge—absolute knowledge (definitional, quantitative), empirical knowledge (based on reasonable inferences from the study of data), and opinions which often masquerade as knowledge, but which are not.

Also in the mix for consideration are various mistakes of reason that often lead astray, with the overriding blunder being that of confirmation bias sometimes supported by: 1) making something out of nothing, 2) making too much out of too little, or 3) simply seeing what one expects to see, but missing a larger and more compelling reality.

Clearly this process journey can be a nebulous endeavor with it involving the search for ideas and concepts that transcend mundane opinion, superstition and general faulty thinking. The importance of this activity moves to the fore particularly on those occasions when an inherited or adopted worldview collides with sense-based data, essentially involving those who hold to a master narrative that runs against the information conveyed by the senses. When this happens, the master narrative becomes unaccountable to other important forms of knowing.

When conflict erupts, it is not uncommon for some to assert that the senses are not to be trusted, and perhaps in some cases this is motivated by efforts to elevate the authority of Scripture. For the record, I would be the first to admit that the senses do often present challenges. There are many examples of empiricism having failed on some level. Perhaps the most prominent example of this sort was the Ptolemaic understanding of a geocentric cosmology. In that situation the senses conveyed that the earth was at the center of everything else, and this due to the feedback received by observing the perceived arcing of the sun, moon and stars around the earth. In short, the senses led observers astray. It wasn't until Copernicus theorized a heliocentric model of cosmology in order to resolve certain anomalies—later aided by strong evidence produced by Galileo—that a whole new paradigm took hold. Simply put, the intricacies and complexities of physical reality do sometimes require modifications of previously held views, resulting in the possibility of paradigm shifts.

In all of this a tacit component plays an important role, with Michael Polanyi eloquently capturing this reality by noting, "We cannot get outside the garment of our own skin." The ubiquitous nature of this actuality applies to all truth claims—ranging from the scientific to the transcendent sacred realm of religious dogma, and is a call for modesty and open-minded humility. Yet, in spite of the tacit element there is something unique about ideas that are grounded in the senses for the simple reason that it affords a form of validation not available to other questing methods. Ultimately, the goal is to develop an interplay which can optimize success in connecting the dots on the map of reality.

Given the nature of this conversation it has been necessary to consider the role, limits and vulnerabilities of sacredly sourced ideas. As suggested above, such a discussion seems important to have because religious constructs can sometimes assume to have something close to an absolute correspondence to reality, with assertions framed as *truths*. When people proceed with absolute certainty regarding the ideas they hold, it can become the pretext for pathological acting out—either physically or emotionally—and thus can be very damaging to the social and institutional fabric. There are many examples of this and some are discussed in the following pages—among these the 9/11 hijackers. So unless the provisional nature of ethereal matters is recognized by the participants it is all too easy for dogmatism to take the driver's seat and then assume a destructive role. The modern world is full of sectarian extremism, and perhaps one of the best antidotes to it would be a more thoroughgoing commitment to a grounded systematic process, recognizing that the substance of a belief can only be credible if the process is credible. After all, there is no essential correspondence between a faith-based belief and the reality, even though many seem inclined to confuse such delineations by making the mistake of equivalence. For this reason, a wise religious tradition must look for ways to integrate or at least abide the distilled wisdom of the ages afforded by science, otherwise the potential increases for falling out of touch with reality.

Some reader may find this discussion regard the sacred uncomfortable. Yet the conversation seems necessary—not to tear down—but to reset how some of these issues are approached. Certainly there has been no intent to suggest that religious presuppositions as a category are inappropriate—simply the importance of understanding their epistemological limitations. One thing is for sure, in spite of the fallibility of the senses they do keep humans grounded in a way that is not otherwise possible. The unfortunate fact is that reality generally gives up its secrets rather grudgingly, and it is through sensitivity to this fact that it may convey the wisdom of modesty in the assertions made. This would seem to be particularly true on ethereal matters for reasons that are developed in the book.

This discussion is submitted respectfully with goodwill intended, in the hopes that it can be a dialogue starter. Being a pragmatist I do anticipate that some readers will initially be uncomfortable with how some sacred ideas are dissected. But for those who endure to the end, it is my confidence that there will be much with which the religious reader can identify, thereby providing a basis for perhaps a more thoughtful and responsive theological worldview—one that is empowered to frame the sacred in a way that is informed by human sense and reason.

Jan M. Long

Preface

It is not uncommon for people to be connected in some way to a faith community. In some cases, such communities operate with the sense that they are God's chosen people. Such thinking is generally sourced in a sacred tradition, and can go to the heart of an individual's worldview. Consider, for example, the Torah (the first five Books of the Old Testament) that tells about a band of nomadic Semites who believed they were chosen. This feeling of specialness has added a powerful dimension to the historic narrative of Jews. Also, consider the Muslim community which teaches that the prophet Mohammed received a series of messages from God, with Muslims the world over now sharing a sense of chosenness as a result. Christians likewise generally have a sense of divine chosenness through the life and message of Jesus, but it is worthy of note that there are also specific Christian denominations that claimed the mantle of "a chosen people." For one, Catholics have a sense of chosenness in their belief that Jesus appointed Peter as the first head of the Church and that this linage traces through to contemporary popes, providing sort of a provenance for such claims. Meanwhile, Mormons believe Joseph Smith had a series of divine encounters resulting in the Book of Mormon and other literary works. These writings play a central role in the life of the Denomination, and for them he is embraced as a prophet. In a similar way Seventh-day Adventists had a nineteenth century visionary founder, Ellen White, whom many of the faith believe was God's messenger. Most Adventists today have a sense of being God's chosen because of their connection to her. In all these examples the sense of chosenness has afforded a powerful cohesive element that other communities sometimes lack.

It is this latter group that I am most acquainted with having been born and reared in a conservative corner of the Midwestern Bible Belt to parents who were members of the Seventh-day Adventist Church. While Adventism has a few distinctive beliefs, in most ways it stands squarely within the mainstream of conservative Christianity. Specifically, in this chapter I will provide some foundation for the conversation that develops in this book by discussing issues that have come up within Adventism that put it into conflict with the dominate scientific interpretations of multiple streams of data regarding the age of the earth, as well as other issues that have gained an elevated level of attention within Adventism in recent years as the Church has drifted formally out of sync with science. But this is not just an Adventist issue as there are many other faith communities holding a similar view, and so in certain important ways this is trans-denominational discussion. Yet, it occurs to me that perhaps the best way to introduce this topic is to provide readers with a sense of how this issue has developed within Adventism. It is a fascinating story in its own right.

In general, Adventists stand squarely within the mainstream of conservative Christianity, but in order to develop this conversation, it may be helpful to detail out a few distinctives. More specifically, the subject of this book—having to do with the relationship of a sacred narrative to the sciences—does not apply exclusively to Adventists.

But for Adventists, the influence of Ellen White has left a profound mark, a woman who was the founding visionary of Adventism, and who left an enormous mark with the content of her visions having been detailed out in numerous books that were intended as guidance to members in how to lead good and upright Christian lives. This included a great many exhortations for how to traverse through a sinful world successfully. Some of the advice was principle-based and timeless, but some of it was clearly a product of a nineteenth century worldview. Ellen White never referred to herself as a prophet, preferring the designation as that of a *messenger*; always holding that she was a lesser light to point the way to the greater light—the Bible. Her influence has been a catalyzing force in an organization that now has a world membership that is approaching 20 million, with a major educational system and a very large network of hospitals and medical clinics that span the globe. It is very doubtful Adventists would have had this success had it not been for the influence of this women.

As a child growing up in an Adventist home I very much lived with the sense of being a part of God's chosen. There was a high sense that God had provided Adventists, through Ellen White, the formula for living the sort of measured life that would eventually bring an eternal reward. Such views were reinforced through church school attendance, as well as weekly church attendance, and as a part of my heritage it afforded my childhood with a sense of security. In short, life was largely centered on family, school and church—with all entities pretty much singing the same tune. It would be my guess that most objective outside observers would likely have regarded this all as a rather insular upbringing, in part due to restricted exposure to other worldviews.

While Adventist thought and practice is quite structured with a great deal of cohesion, it is not monolithic; there are, if fact, many progressive Adventists, though the community in which I was reared did not afford contact with this larger perspective. While it can be debated as to the merits of such a narrow exposure to the world of ideas, I am certain it was not the worst of possible upbringings for among other things I had loving parents who sought what they believed to be the best for their three sons. They were very self-assured in their beliefs and did their best to transmit these values—and thus my childhood proceeded with a high level of cognitive assurance and a sense of security in the knowledge that my tradition had "the truth." Such thinking went a long way towards shaping and molding my early life.

Most of my academic training through graduate school was through Adventist educational institutions, though I did obtain a degree in law outside of the Adventist system. The main point here is that along the way I had repeated exposure to Adventist history and dogma, but to the credit of this education I was also schooled in the methods of critical thinking, and was provided the tools by which to examine the presuppositions and traditions transmitted by birth.

While this book is not about Adventists per se, its inspiration is driven by recent events within the life of the Church. For that reason, it would probably assist readers who are unfamiliar with Adventist thought and practice to step out a few major distinctives that will be foundational to later discussion.

As noted above, in many ways Adventism can fit comfortably within the rubric of Christian fundamentalism though historically the Church has distinguished itself on some issues. One of the important points of differentiation historically has been on the question of inspiration, for unlike most fundamentalists, Adventists eschew the concepts of inerrancy and verbal inspiration of Scripture—at least conceptually. This perspective evolved from firsthand knowledge of the processes employed by Ellen White in writing and publishing the content of her visions. She engaged secretaries that assisted in the writing of her books and articles, yet there was no serious contemplation by most of her contemporaries as to whether she was the architect of these writings. In fact, the founding leaders of Adventism were familiar enough with the processes Ellen White employed to avoid any claim that her messages were verbally inspired or inerrant. It was a rather simple process for them to analogize the vision-to-written-word process they were familiar with to that of the biblical authors. So on the question of inspiration, Adventism did not take the path of fundamentalist Christianity.

In practice, however, this whole subject is a little more complicated for there has long been a tendency of some to elevate both the writings of Ellen White and of Scripture, and treat them as inerrant and verbally inspired. This evidences itself time and again in how they are used, and the level of authority to which they are sometimes elevated. Perhaps one of the most significant results of this tendency has been that many read the Bible very literally and inflexibly, and claim sola scriptura to rule for faith and practice. This same approach has also been evident in the writings of Ellen White with a large number of books having been published posthumously, known colloquially as "the compilations"—publications that comprise many of the large collection of private letters, as well as articles she previously had published that are held in the White Estate Archives. As for the letters directed to private parties, they often consisted of advice on a variety of subjects. These have been arranged topically—with seemingly little thought given to time, place and manner. Her opinion on all topics is considered by many to be sacrosanct, and often carries the import of messages from God universally applicable in detail for those living in the twenty-first century. As such, some are inclined to check their brains at the door; with debates often either started or ended with a good Ellen White quote.

Another distinctive rooted in Adventism is its early tradition of anti-creedalism, often referencing the Bible as being their only creed. This vague doctrinal construction was also very much inclined to treat "truth" as a dynamic enterprise (calling it *present truth*), believing that human understanding evolves. But as the early founders gelled on the primary pillars of the faith, it frequently became a habit of referring to these doctrinal positions as "the truth"—inevitably transitioning from a dynamic process, to one that can be characterized as a static fixation on tradition. This dynamic-to-static transformation has now brought contemporary Adventism to the threshold of creedalism. In recent years the Church has formally hammered out 28 Fundamental Beliefs, and the tendency has been to make acceptance of these beliefs conditional for Church employment; it has also been adopted as an alternative baptismal vow that conditions Church membership on acceptance of these beliefs. In this way, the Church has slowly migrated towards creedalism.

While there are a number of other distinctives, I will mention only one other that is more substantive in nature and it relates to interpretations of the early chapters of Genesis. Ellen White understood Genesis to be presenting a literal description of beginnings. As a result it was her belief that creation of the universe occurred over a 6-day period around six thousand years ago—and because of her visionary history many have assumed her views on this subject to be instructions received from God, though a more likely explanation comes from Bishop Ussher's chronology found in the margins of most King James Versions of the Bible published in the nineteenth century.[1] On this subject, Ellen White understood each day of creation to be as literal 24-hours periods. She also believed there to have been a worldwide flood that drastically altered the face of the earth, and consequently had little use for geologists or other scientists who were, even in the nineteenth century, beginning to offer up a far different narrative of early earth history than what a literal reading of Scripture would allow. Adventists adopted her perspective early on, and have not officially wavered in spite of the mounting evidence that this approach is difficult to justify on scientific grounds.

To its credit, Church officials realized more than fifty years ago that this latter distinctive was becoming the source of increasing tension, so created and funded scientific research through the Geoscience Research Institute (GRI), seeking scientific validation of its historic position. As I understand it, the Institute has spent significant resources looking for weak spots in the scientific data, as well as ways in which to bolster the Adventist hypothesis regarding beginnings. But to date very little data has been found that could offer a more compelling paradigm than what the current scientific model offers. In light of this dilemma, it is ironic that in recent years there has emerged a more militant and intolerant strain of Adventism that is inclined to push out all members who either side with the scientific data, or who urge the Church to take a more nuanced and cautious approach that would allow for a bigger tent of inclusion. This of course has major implications for the Church's educational institutions that have found themselves in the middle of a brewing battle between sectarian orthodoxy, science, and possibly accrediting agencies.

When I started this writing project, some influential elements were calling for a revision of the Church's doctrinal position on creation, a revision that would change its current broad wording into language that is more narrowly drawn and which at its core is both non-biblical and antiscientific, with references to creation as "recent" (code language for a few thousand years ago), and with it occurring over six 24-hour days.[2] In spite of the level of scientific data that points in a different direction, many contemporary Adventists are inclined to submit to Ellen White's nineteenth century views on this and other matters, believing such views to be in perfect concordance

---

1. For those unfamiliar with Ussher's Chronology, Bishop James Ussher was a 17th century Irish cleric who went through the biblical chronologies and concluded that the creation of the world occurred in 4004 BC

2. In using the term non-biblical, I am merely reflecting the fact that the Bible itself does not define the age of either the earth or the universe. As for being anti-scientific this will be discussed in more detail below, but any position that contradicts data, such as inserting the word "recent" as in a recent creation, is by definition antiscientific.

with reality in all particulars. In July 2015 at the Church's General Conference session in San Antonio, Texas, this narrow view was elevated to formal status in a revision to the increasingly creedal statement of Fundamental Beliefs.[3]

In thinking about this turn of events, it has become necessary to ask the question: "What is the appropriate relationship that should exist between science and religion, particularly when conflicts emerge?" In thinking along these lines, Karen Armstrong, who has written extensively on religion, offers her take on this issue. It surfaces in her book on fundamentalism where she provides a working definition of fundamentalism—that being an attempt to merge mythos (myth) with logos (data, reason).[4] She mentions the fact that "higher criticism" is seen as a bogey by Christian fundamentalist who see it as an assault on religion, and who believe there is an accordance between mythos and logos, when in fact such is not possible.[5]

Certainly Armstrong's definition of fundamentalism does not square entirely with my own experience, and I say this, recognizing the enormous contribution she has made toward a fuller understanding of all monotheistic religions. While I would grant that she is probably accurate in so far as the prevalent attitudes fundamentalists hold for higher criticism—seeing it as an enemy to be fought, it is the rest of her statement that seems problematic involving the merging of mythos and logos. It is here I consider her to have seriously missed the mark. For one thing, a fundamentalist rejection of higher criticism—a seemingly accurate observation—is an argument on one level against merger rather than for it, since it is mechanically impossible to reject higher criticism and at the same time be supportive of logos. But for the vast majority of the world's inhabitants in my view who approach religion on fundamentalist terms, dogma is believed by adherents to have a certain correspondence to reality—and this is something not significantly different from those who do not identify as fundamentalist. So, contrary to Armstrong's view that fundamentalists seek to merge mythos and logos, it is arguable that all sane faith communities see their belief system as having a correspondence with reality, and therefore normative—so this is not just a fundamentalist approach.

The real point of distinction for fundamentalism from other faith communities in my view has to do with what sources will be accepted as authoritative, as well as the degree to which they are willing to commit to the notion that they have cornered the market on *truth*.

---

3. For a discussion of changes to the Fundamental Beliefs, please refer to Appendix B
4. See Armstrong, *The Battle for God: A History of Fundamentalism*, pp. 95–96
5. Ibid., 95–96

PREFACE   xix

When a faith community pronounces that they proceed on the basis of *sola scriptura*, it may sometimes signal resistance to outside data that might otherwise influence an interpretation of a sacred text. This can leave the fundamentalist in the unenviable position of holding that a private source of information (such as an interpretation of a sacred text) is more connected to reality than data accessible to the senses. In many cases this would seem to be a function of schooling and intellectual sophistication, while in other cases it seems to be willful blindness.

By way of illustration, in the case of the Geoscience Research Institute (GRI), the goal as observed from the sidelines has not really been a search for truth, for the Church has believed all along it already possessed the correct understanding as to the *how* and *when* of creation. The perhaps unstated goal of GRI has been to find data that supports the Church's theology, while minimizing or explaining away problematic data that is contrary to the traditional narrative. So, in light of this, I am inclined to conclude that perhaps a better way of understanding fundamentalism is in terms of methodology and what it is willing to count as evidence for a worldview.

With the foregoing acting as foundational, I would like to now briefly outline some of the more recent narrative of my own life that motivates this current conversation—namely "young-earth and young-life creationism." Having been reared on the Ellen White understanding that the creation event detailed in Genesis occurred a mere few thousand years ago—such views organically became a part of my spiritual DNA. Genesis 1, of course, is not just talking about the creation of earth, but of the entire universe. This would be the face value conclusion to be drawn when reading that on the fourth day God created the sun, moon and stars. When Ellen White discussed the six-day creation week she was focused primarily on the creation of earth, but there is evidence that she believed the creation of the universe was part of that first week in Genesis. This has been a common view within Adventism until more recent years.[6]

In coming from this background I was often puzzled by scientific dating that suggested a very ancient earth and universe. It was difficult to understand how scientist could be so far off the mark.

I have lived with the mystery of this dichotomy all my life, and because my educational track was not in the sciences I resolved in my early adulthood to enter upon a personal quest to study the evidence and settle this matter to my personal satisfaction—the primary objective being to learn enough science to find peace with my inherited view of beginnings, believing steadfastly they were roughly accurate. But this research did not

6. For further elaboration on this, please see Appendix A

go as planned for the more I studied, the more apparent it became that the scientific data and methods were generally sound and that my understandings regarding the creation of the cosmos as having occurred a mere few thousand years ago would require significant modification.

For example, distances within the cosmos are so vast they are measured in light-years—this being the maximum distance light can travel in the span of a year, yet many galaxies are observed at billions of light years distant. Thus, it is impossible to argue that these distant stars came into existence a mere few thousand years ago since the light seen from those distant galaxies most certainly started its journey to earth billions of years ago. If these galaxies were created a mere few thousand years ago, they might be there in the sky, but no one would see the light from them due to the light-speed limit. Likewise, it is difficult to make intelligent arguments that would deny the validity of radiometric dating. Some do, and I sort of grew up in the Adventist culture that promoted the bogus and unreliable nature of such methods, yet once digging into the science behind it, it is quite clearly a sound methodology. A short chronology of earth history must also confront the vast age evident from ocean sedimentation, the geologic column, ice core data, plate tectonics and continental drift. All point unambiguously towards a very old earth.

Adding to some of this intellectual turmoil within Adventism were discoveries that exploded onto the scene in the late 1970s and early 1980s over the humanness of Ellen White, with revelations that she had borrowed extensively from other authors in many of her signature books without attribution. Some critics were quick to overlook her very limited formal education that might have otherwise led to more charitable conclusions on this matter, given that "truth" is not part of a private domain. Nevertheless, it became a point of crisis for many who had previously understood her gifts to be more strictly of a transcendent nature. Added to this was medical analysis that her visionary states may have been triggered by temporal lobe epilepsy (TLE) that could have been brought on by a well-known childhood head injury that had left her in a coma for a number of weeks. Though this prospect seems plausible, a number of credible defenders were quick to point out that God has the capacity to use all sorts of situations in achieving his purposes.

Those who have been more open to reading Ellen White discriminatingly have been inclined to differentiate ordinary statements (where opinion is more likely), from her "I was shown" type of pronouncements where she was referring to specific details of her visionary states. Traditionally, most members have regarded such statements as beyond critique. Yet, even here, there are problems. Consider, for example, her statement in 1856 regarding attendees at a conference, many of whom she thought would be alive to see the return of Jesus. In describing the content of a vision she had that related to this conference, she makes the following comment:

> "I was shown the company present at the Conference. Said the angel: 'some food for worms . . . some will be alive and remain upon the earth to be translated at the coming of Jesus.'" [7]

Does the obvious error of this statement falsify her entire ministry? Well, I am reminded that a statement is attributed to Jesus in which he predicted that his return to earth in the second coming would occur within the generation of many of those then currently alive.[8] If anyone cares to do a Google search on some of the pronouncements of Joseph Smith, Mohammad, or other oracles, it will be discovered that this same sort of problem also exist in other faith traditions, and is probably somewhat universal to all "revealed" source material to a greater or lesser extent. The obvious cautionary tale from all such situations would seem to be that even when a person believes in and is committed to a prophetic source, it is prudent to recognize the humanity of the source, for there is no evidence that any of these revelatory sources got it completely right.

The factual proof for Ellen White's humanness should be a tipoff as to the risks involved in allowing her authority on Genesis-related-matters to eclipse that of 21st century science. Yet as noted earlier, delegates to the Adventist Church's world convention in July 2015, apparently throwing caution to the wind have now voted to elevate extra-biblical concepts of young earth creationism to the status of official dogma. This latter action has been a major wakeup call for many, acting as the leading edge of a personal crisis for Adventists who find this institutional drift difficult to support.

At this juncture it is worth noting that this is not just an Adventist problem, as a young earth, young life perspective of Genesis extends far beyond the 20 million Adventists in the world. A 2014 Gallup Poll indicated that 26% of Americans believe that humans came into existence less than 10,000 years ago, and a much larger 37% believe the world was created in six 24-hour days. Those percentages represent around 83 million and 118 million people respectively.[9] These are not insignificant numbers, representing ideas that clearly resonate for a very large swath of the population. For this reason, anyone who values both the profound contributions science has

---

7. See White, *Testimonies for the Church*, Vol. 1, 131–132
8. See Matt 24:32–34; Mark 9:1
9. See generally Saletan William, "God's Work?" *Slate* http://www.slate.com/articles/health_and_science/human_nature/2014/12/creationism_poll_how_many_americans_believe_the_bible_is_literal_inerrant.html

made to the world, as well as well as the importance attached to faith, should have motivation for resolve conflicts in this area. This book is intent upon speaking to those who seek to understand the reality rather than blindly serving some narrow parochial interest.

In pondering the staggering numbers of individuals who read Genesis on very literal terms—terms that put it in conflict with the most reasonable interpretations of multiple streams of physical data—it is evident that this subject is in need of careful study. With this in mind, a thoughtful analysis becomes the central focus in the pages to follow. As such, it is not intended as a sectarian response focused on Adventism specifically, but will represent something that will have broad application for anyone connected with a sacred text tradition.

A number of years ago the late astrophysicist, Carl Sagan, wrote a book entitled *"The Demon-Haunted World: Science as a Candle in the Dark."* In it, he encouraged readers to learn critical thinking skill—skills that hold ideas up to rigorous questioning and skeptical thinking. He made the case for peeling back the layers of ignorance and superstition by fully embracing the values of science. While his argument has appeal, it is important to be mindful of the balance needed on this point for the pre-logical and a-critical nature of all inherited worldviews means that no one can completely escape superstition. Nevertheless, there is a well-defined process that has been developed for attaining knowledge, and one of the tangible proofs it is on the right track resides in the astounding success it has had.

The application of process thinking to theological matters will move some readers outside their comfort zone, because among other things it requires thinking to be nimble and tentative, yet anyone interested in optimizing contact with reality will have no choice but to stay fastidiously connected to this defined process.

Those inclined toward categorical thinking and speaking may fail to fully understand the distinction between *personal conviction* and what is meant by the term *knowledge*. The goal is to facilitate discernment as to when modesty is called for in the assertions made. This would seem to be particularly true on ethereal matters for reasons that will be developed in the pages to come.

Because of the sensibilities associated with a more measured approach, it may be a bit puzzling that categorical thinking is so prevalent. While I would not pretend to offer up the last word on this, it would seem that in light of the human condition those who infuse a worldview with expressions of certainty do so largely as a human response to the general insecurities of life. It may be that generalized anxieties are influential in the building of intellectual and spiritual edifices that give structure and support. If so, then they may in fact serve a purpose even if woefully inadequate as a reliable method of staying connected to the real world.

In the first chapter consideration will be given to the whole process of knowing. In fact, much of this book will marinate in this area of thought, including the structure that has been built around the process. After all, a valid faith is an affirmation of something that cannot be fully supported by data, but never something contrary to data. The understandings that emerge from the processes of knowing can inform faith and help guide it when under assault, and thereby keep it on a self-correcting path.

Jan M. Long
Riverside, California

Introduction

> By all outward appearances our life is a spark of light between one eternal darkness and another. Nor is the interval between these two nights an unclouded day, for the more we are able to feel pleasure, the more we are vulnerable to pain—and, whether in background or foreground, the pain is always with us. We have been accustomed to make this existence worthwhile by the belief that there is more than the outward appearance—that we live for a future beyond this life here. For the outward appearance does not seem to make sense. If living is to end in pain, incompleteness, and nothingness, it seems a cruel and futile experience for beings who are born to reason, hope, create, and love. Man, as a being of sense, wants his life to make sense, and he has found it hard to believe that it does so unless there is more than what he senses—unless there is an eternal order and an eternal life behind the uncertain and momentary experience of life-and-death.[10]

In spite of the sublimity of life, it is difficult to ignore the great transitory shadow that hangs over it. As the excerpt above so clearly articulates the human dilemma on the face of it appears to be rather grim, particularly in contrast to those who follow a faith tradition that may inspire a transcendent hope about the future. In the latter case, some of the existential angst can be largely avoided—at least until discovery should be made that some part of that tradition collides squarely with sense and reason. On such occasions a crisis can develop for those who are ill-equipped to deal with such dissonance. One path may lead a person to lose a connectedness to the real world—this due to the powerful tug to keep traditions steady and unchanged. Others may find sense and reason as speaking so powerfully that sacred traditions are abandoned entirely. But perhaps there is a middle ground that is worth exploring, with this at least being the main purpose of

---

10. Watt, *The Wisdom of Insecurity*, Pantheon Books, (1951), 13

this book. To this end I would propose that it is the process elements that go into thinking that is critical to coming to terms with the nature of reality.

In the face of dissonance between a sacred construct and real world data, perhaps the place to start is to take note of the strategic relationship of humanity to the universe. From a sense based footing, earth represents a miniscule spot of real estate in a cosmos that measures billions of light years across. In fact, it is hard to disregard the rather profound insignificance of sentient beings in the grand scheme of things afforded by outward appearance. Yet in spite of this sobering reality a great many people maintain a deep reservoir of hope regarding cosmic relevance. Not only is *relevance* a core longing, but there exist a variety of narratives that support this idea. These are often developed through religious dogma that elevates human value and significance, and it has played an important role in the evolution of human consciousness. In all of this, it is rather common to think of these grand narratives as being reasonably connected to reality, even though a host of factors can intervene to create biases and ultimately cloud judgment—particularly on those occasions when new data calls into question the legitimacy of some aspect of a held worldview.

As meaning-making creatures, humans seek ways to make sense out of the human plight—one that includes a great deal of insecurity. First and foremost, the observation can be made that the natural order seems to proceed with indifference to life, it having the capacity to indiscriminately dislocate, maim, or kill. In addition to this, there exist dangers lurking at the hands of those who operate with hostile and predatory intent. The November 2015 terrorist attack in Paris, France stands as a prime example of this very point. Beyond this, however, there are countless other ways in which life can be compromised, including parasitism, infections, cancers, and harmful chance events such as accidents—to name a few. In addition to all these risks, it is worthy to note that it is not a level playing field, as some deal with hardship and adversity that far exceed that with which most have to deal. On top of all these inherent and tangible threats are the insecurities that attach to the shifting sands of the social, economic, and political landscapes. It leads naturally to questions of the seeming pointlessness of it all. Such thoughts can truly lead to moments of anxiety and despair for purpose-seeking beings that want desperately for it all to matter—to be part of some grand purpose.

Historically, humans have dealt with these enumerated physical insecurities by developing a number of creative strategies to improve survival prospects. The sciences have played a prominent role by extending and vastly improving the quality of life through the management and sometimes cure of disease; developing the ability to preclude many weather related disasters through prediction, warning, and the engineering of more secure shelters. Strategies of offense and defense have also been key in keeping at bay the actions of others—actions that flow from greed, fear, hate, tribalism and barbarism. China built the Great Wall and Europe of antiquity built fortified cities, with much more sophisticated methods used in the modern world. Not to be forgotten is a body of law accompanied by enforcers of the law, designed to regulate society and to protect it from the nefarious and predatory.

To some extent all these creative endeavors have contributed to freeing humans from potential harms to which they might otherwise be exposed; yet the ontological reality continues to give reasons to fret since the myriad of risks that still lurk in our world can never be fully tamed; most of all is the inability to escape mortality. Beneath the surface of these ongoing insecurities are a series of "why" questions that ultimate become part of the domain of religion—which attempts to tap into something that extends not only beyond self, but also science. Through narrative, hope can be inferred that divine control will protect and delivery the faithful from many difficult circumstances of life; that the arrow of reality will eventually lead to a superior plane of existence commonly referred to in monotheism as *heaven*. This narrative can be the source for a purposeful and meaningful existence.

Yet in thinking about this general narrative it is important to remain mindful of the limits to human knowing, and this is itself a part of the human dilemma. While legions assume religious narrative to be above critique, there are reasons for a more sober assessment, with it framing the human context and place in the cosmos through powerful and influential stories about the universe that all come with certain strategic limitations. My friend, David Wilbur, states a truism regarding the limits of narrative in the following: "like all human stories, [they] are incomplete—never broad enough or deep enough to represent the complex and changing universe. They are also distorted by the human needs of those who construct them and those who retell them."[11] In short, he is proposing there to be an important human component in all of this. Such a possibility will be considered in more depth in chapters to come, particularly because there is often a tendency for people of faith to make excessive claims about both the source and authority of religious narrative.

For now, however, consider briefly the fact that more than half of the world's population identifies with one of the triad of monotheistic faith communities (belief in one God)—Judaism, Christianity, and Islam—with

---

11. Wilbur, *Power and Illusion*, 1

the central narrative of each containing the idea that God has revealed his will to humanity. Yet each of these traditions present differing portraits of deity, and unfortunately there are no completely objective grounds for determining which among these variant religious traditions comes closest to expressing the divine character. Thus, humans by nature frequently identify with a tradition and its narrative either as an accident of birth and culture, or by finding a given narrative that has cognitive resonance that then leads the individual to embrace it by ordering life accordingly.

In the end, it all comes down to narrative. Regardless of whether or not a person subscribes to theism (that is, the belief in God), humans either consciously or unconsciously are *narrative* driven beings who seek to explain reality in a way that makes sense. Narrative forms around an inherited worldview built on a lifetime of experiences. As such, it can involve a process by which a person sees, hears, or experiences something, and then tells a story to explain it by integrating it into a matrix of understanding.[12]

Narrative often kicks in to bridge some level of ignorance, and through it, events gets contextualized, impacting a person in ways both positive and negative—depending upon how the narrative holder chooses to evolve it. Narratives contain all the foibles of humanness and therefore will always represent a subjective account. Yet they routinely become the holder's understanding of reality, getting elevated to a working description of how things really are. Unfortunately, some are prepared to defend these subjective understandings—right or wrong—to the death.

Because of reliance upon narrative to such a significant degree, it may be useful to have some clarity about the two distinct sources from which it gets constructed—*logos* and *mythos*. The elements of *logos* include data that comes from observations, and from it enables the creation of a comprehensible matrix of understanding grounded in the senses. As such, narrative has the capacity to be based in the real world. But logos often presents gaps, and these gaps are many times filled in by *mythos*—the elusive elements of subjectivity, including preconceptions, traditions, and constructs of faith, all of which become explanatory filler for the gaps in knowledge that are not readily amenable to examination by empirical methods. Wilbur again offers up some provocative thoughts when he says,

> Many ideologies make sweeping and often complex claims about the nature of reality. These are mostly based on untestable assertions—either about the supernatural, the distant and poorly accessible past or some private source. Once enmeshed

---

12. See generally, Patterson, et al, *Crucial Conversations*, McGraw-Hill Education (2011)

in a human group their complexity, wide acceptance in the group, and desirability keep most people from examining them critically. They then function as powerful placebos. The world is interpreted in their light, people feel good about that interpretation which reinforces their belief and other more rational understandings are rejected as deficient or corrupt.[13]

Those in the faith community are many times offended when the term *mythos* or its derivation is applied to belief, and this is likely as the result of assumptions that the idiom suggests a lack of correspondence to reality, even though the more appropriate way to think about it is that it merely refers to ideas or beliefs that do not reside on a sense-based footing. As such, there simply cannot be laboratory testing for most faith propositions. In such cases humans are often limited to accept, or not, constructs that evolved from the distant past, from which a worldview has been inherited. Being mindful of the multitude of variations associated with religious notions of one sort or another floating around in the marketplace of ideas, there should be an instinctive sense that they cannot all be right. The term *mythos* or its derivation *mythology*, then, becomes a stand-in term for attempts at objectivity by putting all these variations into one category. The common inclination is to attribute the status of *myth* in its more negative association to other people's beliefs, but to exclude one's own. In this latter case, it is easy to mistakenly attribute it to having a pristine concordance with reality, which in sectarian circles is usually termed *truth*.

In general, most recognize that many aspects of religious belief are somewhere beyond the reach of the empirical world, yet it is important to understand that this is not always the case—the early chapters of Genesis being a case in point. Yet because some view data as a threat to a standing narrative such as this, it is easy to understand why some oppose any process that will put old ideas in jeopardy. This, then, can become the juncture at which the narrative holder decides whether mythos or logos will rule. The alternative is to find some form of unifying approach between the two. These can be pain-filled and stressful moments since religion very often provides the context for how humans see their place in the world, and any narrative modifiers can cut deeply into personal identity. [14]

However, it is possible to minimize the stress of it all if it is recognized from the outset that one's understandings of reality are feeble at best. It certainly represents a more humble expression of faith—with the practical

---

13. David Wilbur in a presentation made to the Centennial Seminar Group, Loma Linda University, March 23, 2013

14. see for example, http://www.answersingenesis.org

recognition that finite humans "see through a glass darkly." Indeed, it may be helpful to consider the possibility that some faith constructs are sourced in human desires and longings—either by those who hold them now or those who originally gave birth to the idea. By thinking in these terms, it at least fosters an attitude essential to the exploration of what will unfold in the following pages, for unless the sources and limits of narrative are analyzed it increases the likelihood that attributions will be made that lack credibility on some level. Humans have the capacity to keep it credible by using sense-data and reason as a check on the claims sometimes made that are not so based.

Occasionally, data may upend a meaning-making narrative—including specific beliefs and assumption—doing so with sometimes devastating consequences by leaving its holder unmoored. In the words of Karen Armstrong, "unless we find some pattern or significance in our lives, we fall very easily into despair."[15] At the very least there should be some anticipation of an unsettling effect when discovery is made that what was previously assumed to have a certain correspondence to reality is either wrong, or is in some way inadequate. The alternative is to live in denial of human sense and reason. It is on this basis that there may be wisdom in approaching narrative with a more open approach, it offering the possibility of modifying a narrative in a way that can perhaps avoid the psychological trauma that might otherwise occur.

It may be helpful to briefly consider an example from history when new data resulted in a crisis for many individuals. Probably the most vivid illustration of this sort emerged from the pre-Copernican view of earth being the center of the universe—often referred to as geocentrism—where the dominant understanding was that earth was at the center of the universe, with the sun, moon and stars all revolving around it. For one, that was the obvious appearance of things when looking up into the sky. While this was the scientific view, it was also etched into the religious narrative of the age, which described in rather egocentric terms human importance in the whole cosmic scheme of things. Geocentrism acted as "Exhibit A" in this regard. But as is now known, that view was unsustainable with the accumulation of later acquired data that exposed certain inadequacies with the model. This then led to a new more satisfying model that encompassed what came to be known as the *Copernican Revolution*. In the midst of this affair there were two distinct groups of individuals—those whose views were shaped primarily or exclusively from the mythos of religious text and teaching,

---

15. Armstrong, *The Bible*, 1; see also *The Battle for God: A History of Fundamentalism*, 135, where this idea is similarly expressed.

and those who reached their views based on data. The resulting crisis impacted both groups, but for those committed to data, and particularly the newly acquired data, it is possible to guess that it was easier to assimilate the adjustment than for those who incorporated their views from a parochial interpretation of Scripture. In time, even those reliant upon religious text found it necessary to adjust previous understandings if there was going to be any hope of being relevant. While some may mistakenly attribute the pre-Copernican view entirely to mythos,[16] it must be remembered that good scientists of the age were actually working with data—just misinterpreting it—so the pre-Copernican view was not entirely myth based.

At the beginning of this section was a quote from philosopher Alan Watts. Those who may be familiar with Watts' writings will know that he was not intentionally attempting to elevate the theological enterprise with his words, yet he powerfully articulates the exact reason why it is an endeavor of significance to humans—it having the capacity to go places that science alone is unable to go. But more importantly it highlights the limits of science, for certainly it is ill equipped to considering some of the philosophical and theological aspects of the human predicament—the sweetness of life enmeshed with great pain, adversity, and suffering. Given the human predicament, it seems that the quest for purpose can play a central role in the pursuit of a meaningful life.

I have spent the better part of my life pondering this mystery, and since I am most familiar with the Judeo-Christian version of the "grand narrative" that attempts to respond to this mystery, I will focus primarily on it; however, I do not pretend to suggest that this discussion is limited to the Judeo-Christian tradition, for it surely has certain universal application.

Undoubtedly one of the best ways to understand ideas that are being presented is to have some frame of reference, and thus the discussion began in the Preface by taking a look at my journey away from fundamentalism where dogma narrative trumps data. Looking forward, the book begins in Chapter 1, with, the whole process of "knowing" considered. It is here that much of the knowledge that is had about the universe is built on a growing body of data accessible to human sense and logic. Modern life is a bold testament to its successes, acknowledging that in its shadow faith has not always faired so well. The idolatrous manifestations of faith often become all too transparent, with cherished interpretations of sacred writings sitting arrogantly on lofty pedestals far above all other human wisdom. In this regard, Galileo forever stands as the poster-child for the role science has played in

---

16. For example, Armstrong, referenced in note #3, lumps the pre-Copernican view in one category, namely *mythos*.

deconstructing some of the sacred tenets of faith, exposing both error and inadequacy in its modus operandi. This should not be seen as a negation of a role for religion, but what it does is sort of resets the parameters and lines of authority.

From here, there will be a group of chapters that consider possible areas where the religious narrative within the faith community veers off track. It will be one of my arguments that in the context of faith and practice a great many idolatrous constructs emerge that could be labeled *false gods*. Sometimes they are benign, but often they are very destructive. This term "idolatry" is often used in connection with excessive focus on sex, money or power, but is also sometimes applied to excessive and extreme ideological, political or religious commitments. Within the framework of this discussion, the term will be restricted to extreme sectarian commitments—including religious dogma, worship of institutions and iconic figures, as well as unwarranted uses of sacred texts.[17] On these points it is not uncommon at all for individuals to make extra-ordinary claims and commitments that impact others dramatically. At the extreme, there are examples both in history and in the contemporary world of individuals becoming martyrs for "the faith" for what most would deem to be very inappropriate reasons.[18]

In all of this, there may be a naturalistic explanation given that some neuroscientists are now making the case for this view, and it will be important to consider the data. If Scripture represents to some extent the human face of God, this could be an important connecting thread to consider.

Whenever any discussion of faith ensues that seeks a level of objectivity it is necessary to consider the vulnerabilities of faith, that is, those elements that could constitute possible reasons for doubt. Some will view this as dangerous ground, yet if a person of faith has not considered those areas where faith is most vulnerable it becomes a legitimate point of query to ask, "On what basis is an unexamined faith to be counted as worthy of respect?"

Finally, the last few chapters will consider some of the strategies that can be used to salvage the pieces and consider the most credible framing for mythos—this regardless of whether it involves a believer or an agnostic. In order to get to this endpoint, it is necessary to deconstruct some of the common erroneous paths that have evolved.

Some will likely have no interest in this exercise, seeking only comforting conformations of belief—never questioning whether things are on the

---

17. Many little "g" gods are of a purely secular nature—with the most obvious being an inordinate focus on money, power, nationalism, or political ideology. We could spin these out in some detail, but we will restrict the discussion to the more specific god that originates within religion itself.

18. Certainly the 9–11 hijackers would likely be on most lists.

right track in the first place. Yet, those who desire to connect with reality in any compelling way will have no choice but to engage in self-examination. It is the strategic essence of this latter tradition that motivates this project.

Certainly given the influence of religion in the lives of billons of people, as well as its destructive effects in the history of civilization this subject deserves examination. By the end of this exercise it is my goal for the reader—believer or not—to have a better understanding of the demarcation between sense data, reason, and faith. It should also help clarify the true nature of the human dilemma and provide a new path for possible integration of hope into the equation. This is not a definitive literary work by any means, but what I hope to accomplish is to develop a map by which readers can calibrate their lives, to recreate mythologies that govern worldview in a manner that may have a closer correspondence to reality.

The following reflections were outlined by Joe Erwin, a biologist, on his life journey that began with fundamentalist dogma. He posted these thoughts to Spectrum Magazine's website in response to a blog conversation regarding creationism, dated March 14, 2015.[19] It speaks quite eloquently to the dilemma that knowledge can bring to a worldview that is inadequate in its articulation of real world realities.

I was reared in a religious tradition that put high value on an evidence based dynamic faith, one in which truth was always held up as the goal. It was also a religious tradition that interpreted Genesis on very literal terms. This was quite acceptable at a time when I was aware of no evidence to the contrary, but over time I became well aware that this was an area of belief fraught with problems, and resolved to give this subject honest study.

So, what happens when one gives a fair hearing to the sciences—cosmology, biology, and geology?

What happened for me was that I encountered a lot of information that was incompatible with some of the traditions I had been reared with—especially with regard to the age of the earth and life on it.

I also found that many Christians, including some clergy, recognized that geological information was real and compelling, and that many of them saw that scripture contained stories and explanations that could not have

---

19. http://conversation.spectrummagazine.org/t/what-do-we-do-with-creation/7942/77.

been expected to be literally accurate, given the state of knowledge when the stories were told and retold and then finally written down.

I did not seek knowledge because I was skeptical. I simply wanted to examine the facts as objectively as possible—without, necessarily accepting as true either the facts or the explanations. I learned to be skeptical of both facts and explanations, sifting through them and learning to hold evidence lightly. Recognizing that many things are not exactly as they seem. Realizing that I should not grasp any fact or explanation so tightly that I would be devastated if I learned it was untrue.

I guess I could be reasonably comfortable with a creator God who could have done anything, anytime He wished and however He wished to do so; a God who set in motion a complex and remarkable system of life that changes as needed to survive in diverse and changing environments. Surely anyone capable of designing such a system is worthy of admiration, and probably even worship. Spending one's life trying to appreciate when and how all this happened, and even contemplating why, is a fitting tribute to the creator—whether person or process or both.

The trouble with "creation science," along with the beliefs it is designed to defend is that it decides in advance what happened and when, and then, sort of dictates to God what He could have done and when. It puts God in a box defined by humans, their gullible and unsupportable beliefs, and tries to use science to prove things that science has no means of proving one way or the other.

As I age, secure in knowing that I have made a good effort to understand my time and place, to learn how to learn, to seek knowledge but hold it gently, I am at great peace. Things are, however they are. If I am able to understand them a little, I relish that. I have seen much beauty and experienced much pleasure. I am awe struck by all of it—and I have reverence for the remarkable complexity that exists.

In the face of all this wonder, I learn new things every day, with the understanding that I can never fully know or understand anything. But sometimes I think the greatest mystery of all is how anyone is able to close his/her mind around some small array of dogma, and cling ferociously to that in the face of abundant information to the contrary.

I caution against eating from the tree of "Devotion to Willful Ignorance." The consequences can be devastating.

Foundational Considerations

CHAPTER 1

On Knowledge and Certitude

Chapter Summary

THE MAIN PREMISE OF this chapter is that there are rules that can assist in navigating conflicts between the sacred stories that deliver meaning and the sense data that keeps humans grounded to the real world. As such, this chapter sets forth a discussion for those who have a genuine desire to live as connected to reality as possible, providing the intellectual tools needed to traverse past conflicts that can emerge between science and religion, recognizing that some parts of every individual's inherited worldview challenges such connections. The proposal I put forth is that there are good reasons for people of faith to maintain respect for sense-based quests that engage in systematic inquiry, if for no other reason than the fact that history clearly provides guidance on how the zone of overlap should be handled. This book is largely about the search for those principles that can guide a person in the exciting life-long process of getting acquainted with the world in all its incomprehensible permutations. In order to adequately work through the process terrain, it has been necessary to consider some of the major components that go into human thinking—things like the sources of ideas, structure, content, limits and vulnerabilities. It is in this context that consideration is given to differing types of knowledge—absolute knowledge (definitional, quantitative), empirical knowledge (based on reasonable inferences from the study of data), and opinions which often masquerade as knowledge, but which are not. From this foundation, the balance of the book attempts to build on these concepts.

※

As I listen to the broad public discourse taking place today on any number of issues, it seems that there is less concern with actual facts or data, and more concern with holding to a master narrative (or ideology) that may or may not be on the right track. Sometimes such arguments are presented as though there is no other alternative. It is generally apparent when a speaker has skated onto thin ice by thinking and speaking as though understandings are in absolute accord with reality, yet this method of verbal intercourse is pervasive today—particularly on talk-radio and the cable news channels. Such shallowness also infects the faith community where, unfortunately, doctrinaire and uncompromising assertions—truth claims—are frequently made. In this latter case, such attitudes likely track back to religious institutions themselves that sometimes error in projecting an attitude of having a super-concordance between doctrinal points and reality. So what is it about human nature that inspires the spirit of dogmatism and arrogance? More importantly, how is a more mature and defensible modus operandi created? These are pressing questions that civil and religious societies must address today.

In thinking about this matter, I am mindful that all humans operate from a given perspective, and whenever that perspective comes under some challenge, the common reaction is often to search for talking points that can counter the challenge to a cherished understanding. It is normal to have a range of beliefs about any number of things, some being handed down from parents and tradition, while others having been acquired along the way. Some of these beliefs relate to points of significance, while others much less so. They comprise the mythologies that everyone carries, which become the backdrop for the narratives that attempt to make sense of the world, and that weave themselves into the mind. While the general human inclination is to assume a correspondence between a personal belief and the reality, a milestone is achieved upon recognition that it is a lot more complex than what it appears on the surface.

It can be conjectured that what is going on in a lot of cases is an attempt to cover a sort of existential nakedness and vulnerability by projecting with a veneer of certitude in a very uncertain world. Due to existential vulnerabilities, it is arguable that there are motivations to construct as much surety into life as possible. Not only does it surface in common discourse, but also most people purchase insurance to cover the contingencies of disasters, both of health and of material possessions. Locks are put on the door to assure that unplanned guests don't intrude. Money is put in the bank for safekeeping. Mythologies are sometime adopted that fortify the mind with certainty about the nature of reality and the meaning it may hold.

This chapter specifically deals with the certitudes of mind that pertain to knowledge, with the goal being to dissect what we really know. The argument will be that some certitudes are defensible while most are not. Perhaps central to such considerations is the whole question of knowing, for if *knowledge* can be successfully identified then the individual is positioned to determine when and where expressions of certitude are appropriate. This chapter attempts to provide a useful template for reader to apply in their personal lives.

It wasn't until I was along in my own educational development that it began dawning on me that many issues and concepts that previously seemed rather pedestrian and easily resolvable in some definitive way, were actually quite complex, and at deeper levels did not lend well to categorical expressions. Examples abound on many of the social, political and theological issues of contemporary life. As it turns out, articulating a point of view with dogmatic certainty is one thing, but doing so with credibility is a learned skill, and is something that must be honed continually. With that in mind it seems appropriate to review some of the principles that can help move this along.

The most obvious evidence that certitudes of mind are inappropriate in a given situation is when two "certain minds" are locked in disagreement. An outside observer will recognize immediately that both generally cannot be right—and of course there is the chance that neither is right. But unless there be some criteria for making an assessment of the validity of the thinking processes in play of one certain mind verses the other, there will be no basis by which to make an appropriate appraisal other than personal bias.

In recent years I have participated in a voluntary association of interdisciplinary scholars and others who meet weekly in a university setting. This group is engaged in the review of books and topics in the genre of science, philosophy and theology. Those who participate are specialists in many and varied areas of knowledge, and is a setting in which cross learning can take place. One of the principles to be learned from groups of this sort, is that, the more study given to almost any aspect of reality, the less inclination there will be to proceed on categorical terms. In all of this the main distinction between mundane expressions of certitude and a subject-matter expert who speaks with some authority is some level of "knowledge."

Without doing a formal study, it is possible to guess that dogmatic discourse is connected in some way to personal need and narrative—with these factors influencing the receptiveness held towards political ideas, religious dogma and traditions, as well as other grand ideas that attempt to organize one's response to the world. Yet, upon recognizing that the larger reality never precisely correlates with self-perceptions, it can become a watershed evolution of thought. This is key to making appropriate judgments about the categorical essence conveyed in what is thought and said.

The Path to Knowledge

It seems that the key to improved discourse with each other is to develop a clear distinction in understanding as to the processes in play that form the basis for *knowledge*, verses that which constitutes *opinion*. This formalized area of study is known as epistemology and it offers assistance in identifying when expressions of mental certitude are appropriate and when they are not. Strategically, there are several approaches that could be considered. I will mention a few of them, including the following: empiricism, rationalism, objectivism, subjectivism, authoritarianism, and pragmatism. So with the goal of attempting to avoid as much as possible the mire of a technical discussion, it might be useful to give some brief thought to this. In this chapter I would like to offer up an approach that proposes three primary levels in the ladder of knowledge, doing so by combining the senses with reason.[1] Other approaches will be considered in later chapters.

Knowledge in the extreme and absolute sense

Knowledge in the absolute sense tends to be more tedious in its particularity than perhaps other forms, but such particularity affords a structure of self-evident truths about which reasonable people cannot otherwise object. As such, this form of knowledge falls into two categories: 1) definitions and 2) closed systems. As for the first of these, most will be familiar with a type of knowledge that is definitional and that as such possesses a categorical quality formed by the definitions themselves—statements such as a whole being greater than any of its component parts, or a triangle being defined

---

1 I am indebted, here, to the late philosopher Mortimer Adler who discusses epistemology in a number of his books, with his clarity of thought par excellent. For a good representative discussion, see generally, Adler, *Ten Philosophical Mistakes*. I am also indebted to the wisdom of Mailen Kootsey who offered advice that sharpen the accuracy of the details in this section.

as a three-sided polygon composed of three angles, etc. Language is built on definitions, and in this sense it is possible to convey meaning through verbalized sounds (or words) based on a common set of understandings about the meaning of such words. Knowledge of this sort is the product of rational thinking, built on the logical interplay of terms. The second type of extreme knowledge are closed systems. This may include physical systems that do not interact with other physical systems. It can also be thought of as integers in mathematics—2+2=4, or 10+15=25, where categorical meanings have been established. All such evident truths become incorrigible ideas because they cannot ever be understood as erroneous, thereby representing a distinct category in the context of all the assumed wisdom humans have acquired. On those occasions when conversing a point that fit into this classification, it should be recognized that categorical language is most appropriate, but since very little knowledge is of this variety it is a general rule that categorical language is not appropriate.

Knowledge with a practical definition (*sometimes known to be opinion in the strong sense of the word*)

This category relates to conclusions reached based on empirical data and reason—essentially this is the essence of the scientific method. There is an underlying human longing to connect with that which is real and it is generally believed that the closest that can be gotten to reality is through the senses—things that can be observed with the eyes, that can be heard, touched, smelled and/or tasted. Modern instruments act as sense amplifiers to extend the range of the senses, to allow, for example, seeing and hearing that which would otherwise not be possible.

The central importance of the senses themselves comes with the recognition that the only hope for acquiring an approximated contact with reality comes through the data acquired by the senses. To the extent that humans lose their sensate or cognitive capacities through one medical condition or another—contact with reality gets diminished accordingly.

Once data input is acquired, the brain is employed to make sense of it all. It is the end result of this process that is sometimes called *knowledge*. Yet, to be clear, it can never be understood as knowledge in the absolute sense of the word as will be explained below. What is important to understand about this type of knowledge is that it involves the uniting of the senses with the ability to reason, and this has moved humans from superstition where everything was attributed to fate as governed by the gods, to a point in human history where most now recognize that the universe operates on the basis of fundamental laws or principle—from cause to effect. There are, in fact, regularities of nature that can be investigated, from which the laws of nature can be deciphered.

Such opinions often get classified as *knowledge* because they are based on sufficient probative force so as to justify the claim being made at the time that the opinion being rendered has some correspondence to reality. Yet most will recognize from experience that such "knowledge" is sometimes insufficient in the long term. Consider, for example, the case of Newtonian physics. It sets forth principles regarding the laws of motion that have predictive value in describing motion. These principles are still considered valid today, yet Einstein was able to demonstrate that Newtonian physics was inadequate in addressing a broader range of circumstances. The point is, it is impossible to fully escape the subjective realm, even while a practical definition of knowledge can be crafted that elevates sense data and reason—it representing a powerful tool in the human ability to unlock the mysteries of the universe. It can be framed broadly enough to allow for modifications that may be needed at some later point in time as a more complete understanding emerges.

By use of such methods it has been possible to discover that the natural order is governed by regularities that can be observed. What these regularities mean in practical terms is that there are methods for projecting into the future about observed processes in play and these can make reliable predictions about future events (the most obvious being—the sun will rise in the east tomorrow), but also the capacity to trace processes back in time to learn a great deal about the past. These regularities are distilled to laws of nature, and ultimately mathematics. Even quantum mechanics—noted for indeterminacy—is governed by the regularities of probability, much as a random coin toss will yield—over the long term—a 50/50 ratio of heads and tails—even if the next toss of the coin cannot be accurately predicted.

So, while some believers think in terms of magic when reading of some of the Divine actions referenced in Scripture, the regularities discussed above have led others to recognize that perhaps some of these ancient understandings are really the primitive version of what today would be recognized as natural law in operation—law that can be explained by natural means. To acknowledge this possibility allows for biblical narrative to convey poetically certain things that the ancient world could describe in no other way than it did.

The biggest difference between the methods of science and everyday life in guarding against errors of belief is the use of the senses combined with a set of formal procedures to protect against the sources of bias and error. These would include such things as the use of statistics, control groups blind studies, and peer review.

Opinion (*Conclusions not in levels 1 or 2 above should then be regarded as opinion in the weak sense of the word*)

Opinion represents a broad category of discourse that is often short on evidence and/or reason. This class category is well known for shallow logic and cherry picked data designed to reinforce personal prejudice. As negative as this all sounds, it is interesting to note that opinions may be false, but they may also be true. The distinguishing feature is that these are assertions, beliefs, or conclusions that are generally not derived from consistent or detailed use of the senses, or structured thinking. Consequently, this category is fundamentally distinct from levels 1 and 2 above, and the least appropriate for assertions of mental certitude. Yet, ironically, it is often this very category where mental certitudes are most common.

Individuals who are part of a faith community often commit to and defend dogma that clearly does not fit within the traditional definition of *knowledge*, even though the word *truth* is often invoked, as are constructs of certitude. So the question arises as to how faith should mesh with this understanding of knowledge? After all, the matter of religious "truth" is tied up with propositions that are asserted as true, and are generally backed by some form of revelation that is seldom verifiable in a scientific sense. In spite of this reality it is not uncommon for faith assertions to be treated in very concrete ways, often protected by the guardians of orthodoxy who are confident that the narrative they defend represents a collection of unalterable static truths. I will be arguing that even if truth has an absolute and static quality about it, certainly the human component requires that it be thought of more as a process. It is particularly ironic that attitudes of dogmatism very often surround ideas that are among the most speculative on the spectrum of knowledge, and those in doubt should visit an online religious discussion forum. The alternative to codifying the speculative, then, is to treat these more sublime matters respectfully and openly as part of a dynamic and ongoing quest to understand reality.

For many believers the transcendent reality to which faith speaks is very real to the subjective mind of the holder, and yet how is the substance behind the belief ascertained? If the argument is made that knowing comes through revelation, it is important to recognize that there are many claims to revelation and unfortunately they are not all saying the same thing. Because of this fact, any desire to take revelation seriously raises the query as to what criteria should be employed in assess which revelations are authentic and which are not? This is, in fact, a significant problem.

For example, let's consider two types of religious fundamentalists—those who are Christian and holds Scripture to be the inerrant and the literal word of God, and those whose main narrative comes from the Islamic tradition, a tradition that can hold a similar view about the Qur'an. Certainly anyone identifying with one of these religious traditions holding its sacred text to a standard that is not open to outside data (as demanded by the historical-critical method),[2] must recognize that these beliefs will be understood as part of the domain of opinion by those outside the tradition. After all, how is it possible to engage in meaningful conversation with an individual from a different sacred text tradition who may treat his own sacred Scriptures in a similar fashion? Unless a way can be found to achieve common ground through objective formalities, belief is reduced to *opinion* in the low sense of the word. Some readers will not like this characterization, yet this seems to be the appropriate way to describe ideas that do not appeal to, or that may sometimes even eschew empirical data that would be outside the text itself. The fact is if there is any desire to live in the real world there really is no clear way of achieving this short of reference to elements that share a common and universal standard of access.

The DIKW Pyramid

As sort of a summary of what has been discussed in this chapter so far, I would like to refer readers to The DIKW Pyramid which was introduced by Russell Ackoff in 1988 at a professional conference, and was then later published in article form in 1989, with DIKW being an acronym for Data, Information, Knowledge, and Wisdom.[3] This pyramid of knowledge has a hierarchy to it that begins with data at the base of the pyramid, then from data comes information, with knowledge evolving from information, reserving the top level of the pyramid for Wisdom. This model has been critiqued as being too simplistic, yet Ackoff was a professor of management science at the Wharton School, and specialized in operations research and organizational theory. As such this model offers up something of a practical approach, as opposed to something more comprehensive and complex. As a pragmatic template for working with information, it offers a high-level strategic approach for understanding how knowledge is acquired and integrated, and for this reason the DIKW Pyramid has been noted to be "canonical to the field of knowledge organization." Some of this has already been

---

2. Sola Scriptura comes to mind here, coming from the more conservative Christian tradition.

3. Rowley and Hartley, *Organizing Knowledge*, 5–6

covered in the discussion above, but there might be some value in considering each of these categories as it is applied in the DIKW model.[4]

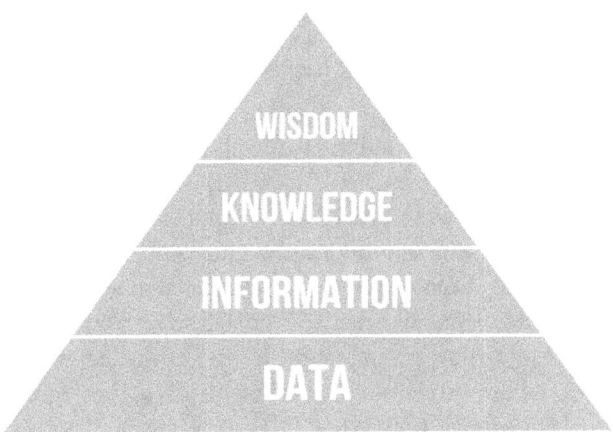

Beginning at the bottom with data, in the sciences it is generally thought of as some type of stimuli represented by signals, symbols or signs, that are meaningless until an investigator is able to put them into some hierarchy of meaning.[5] There are a number of ways in which stimuli gets elevated to "data." This can occur when a sensate being makes an observation that can be recorded in some manner, having gotten an investigators attention. It may be numerical, by video, by diagram, or detailed by written words. One of the early formulators of this model characterized data as being "discrete, objective facts or observations, which are unorganized and unprocessed and therefore have no meaning or value because of lack of context and interpretation."[6] This implies there to be a subjective element involved in the acquisition of data, something that is widely acknowledged within academia.

Meanwhile, information is derived inferentially from data that has been organized into some form of intelligibility by answering the "who," "what," "when," type of questions. In this way, information becomes more than just data, and thus is understood at a higher level in that information classifies and arranges data in a way that affords some meaning.[7]

The elusive and subjective constituent in all of this is the interpretive element involving data and an observer. One definition of empiricism that

---

4. This is the Wikipedia representation of the DIKW pyramid.
5. Zin, "Conceptual Approaches to Defining Data, Information, and Knowledge," 479–493.
6. Ibid., note 3, 5–6.
7. Ibid.

conveys something of its essence, "is that it is a fluid mix of framed experience, values, contextual information, expert insight and grounded intuition that provides an environment and framework for evaluating and incorporating new experiences and information. It originates and is applied in the minds of knowers."[8] It can be understood in a processed context as a mental structure where "information is connected in relationships."[9] Yet it can also be thought of procedurally—involving experience, skill, expertise or capability. Collectively when discussion turns to scientific knowledge, there will always be a significant procedural component involved in as much as process hovers in the background, and is ultimately the key to credibility.

It has already been pointed out that in most traditional understandings of knowledge, religious belief does not fair very well. There will be more discussion of this below, given that the term "knowledge" is sometimes extended to a cognitive framework pertaining to presuppositional categories. It is significant enough to the overall thesis of this book to warrant some further consideration.

Wisdom is the most controversial part of this pyramid, with it variously being described as the "ability to increase effectiveness," that involves making sound judgment, using knowledge for the greater-good, or for some added value. A lot of these terms are nebulous and controversial, though they may be of some pragmatic value.[10]

The Status of Religious Belief

Going back now to the general theme of this chapter, the discussion has focused primarily on knowledge that accrues from data and physical evidence. Those who have very strong religious convictions, who perhaps have always assumed that belief had a correspondence to reality will likely be surprised at the very marginal position reserved for it in the schematic of "knowledge" outlined above. Yet with the outsized success of empiricism, it seems that religious dogma has found it difficult to compete. The mistake often made is to confuse certitudes of mind and personal convictions with knowledge in this more formalized sense. The seduction of mind is to assume that, that which is believed has something on the order of an exacting concordance with reality. Realistically, the most that can be anticipated is a vague correspondence, and the prima facie evidence for this conclusion is found in the innumerable number of religious denominations and doctrinal

---

8. Wallace, "Knowledge Management," 1–14
9. Ibid.
10. Rowley and Hartley, *Organizing Knowledge*, 5–6.

constructs—all asserted as *truth,* even though they, in many cases, assert mutually exclusive dogmas. It is due to this very reality that sacred constructs are often put into the category of mythos.

In an attempt to add to the clarity that mythology is part of all religious systems of thought, it may be useful to briefly retrace the ground that has been covered in this chapter, namely the fact that knowledge is based on self-evident truths, and on information derived from sense data as moderated by reason. Religious beliefs can contain reason, but falls short on empirical data in most cases. Finite beings simply do not have the ability to bring to religious belief the level of certitude that exists in the scientific realm. If this is considered objectively it is not difficult to understand why. This was very briefly mentioned in the Introduction, but if thought is given to all of the many claims that have been made to "revelation" by Islam, Christianity, Judaism, and many others, it has a certain transparency about it. The question becomes: "By what criteria does a finite human determine authenticity of that which purports to represent "sacred knowledge?" Is it simply enough to embrace the religious beliefs of the dominant culture or subculture on faith, or appeal to a text purporting to be of transcendent origin by some tradition without ever considering the basis of the claim. If a sacred tradition or text is deemed inspired, what is the process by which it is deemed as such?

If every sacred text were saying the same thing, life would be quite simple. But since this is not the case humans are left to choose between religious traditions that identify reality in very different ways. So while humans desire certitude regarding the large questions of life, the reality is left uncertain on many levels. The human response to such uncertainties can lead to dogmatism, subordinating cognitive endeavors to sectarian formulas that artificially elevate certainty. In the final analysis finite beings are left no choice but to recognize the very tentative nature of certainty afforded by all religious systems.

One of the things I have learned in my few decades on earth is that there is a tendency to not give critical scrutiny to the ideas that have been inherited. From this uncritically examined perch it is all too easy to incorporate findings that agree with those things already believed and summarily dismiss everything that does not have a nice fit. It is only as the individual steps outside their comfort zone and critiques their presuppositions by giving fair assessment to that which is accessible to the human senses, including data that may run against an operating narrative that it is possible to move forward on the path of knowledge.

Frankly no human knows very much for sure, even when the brain is filled with a lot of subjective certitudes. So in giving study to the framework by which knowledge is acquired, the discovery that humans know less than they may have subjectively assumed should be humbling. But it also stands as an opportunity to overthrow the mantle of dogmatism for a more defensible level of living.

Up to this point I have focused mostly on the role empiricism has had in expanding human knowledge, briefly considering some of the ways this quest can be hindered. But it will be important to consider in some detail the role that reason plays in this whole dynamic, as well as it implications for the sacred realm, and this will be taken up in a later chapter.

Summary Points

There is a lot more that could be said on this subject, particularly on the interface of religious belief and knowledge, and it will be developed in due course. But for now, this seem to be a good breaking point for this current discussion.

So, what are the takeaways from this discussion? Let me propose the following:

1. The way in which understanding of the world is organized (personal narrative) should not be confused with the reality. It is never the same thing even though it may seem like it is.

2. *Knowledge* in the strong use of the term is something of which no one can really claim to have much of. This is not only an ontological reality, but also a posture of humility that can actually be helpful to the quest to understand reality in all its permutation;

3. *Knowledge* in the weak sense of the term is acquired through sense data and reason, and is something everyone has the ability to develop. In essence it represents the foundations for the scientific method;

4. *Opinions* are loosely constructed ideas that fill the mind, and consequently often come tumbling out in normal discourse. People have opinions about most everything, and these represent positions held that may be based on pieces of data, but which lacks any systematic analysis and therefore fails to rise to the level of rigor required for the category of knowledge. Often times it includes personal bias and prejudice, and can have connections to religious dogma;

5. Inferred in this discussion is the idea that error, dogmatism and possible self-delusion can be avoided or minimized by maintaining a sense of propriety and humility about what one truly knows.

6. Each of these points can contribute to a more circumspect relationship to human understanding of the sacred.

CHAPTER 2

The Madness of Crowds

Chapter Summary

The way in which people see the world tends to run the spectrum from tentative and open, to closed and rigid. While for some, evidence is an important part of the equation, others allow traditional ideas to play a leading role. One thing is for certain, though, when a master narrative is out of touch with a preponderating body of evidence, just about any idea can be seen as tenable. On top of that, such ideas can appear credible and more easily accepted when others are participating in the error. Some of the leading processes that contribute to people embracing pseudo-knowledge would be the following: 1) making something out of nothing, 2) making too much out of too little, and 3) the mistake of seeing what the observer wants to see, even while missing other important clues. It is by avoiding dogmatic thinking, allowing sense and reason to play a prominent role, and giving modesty to the declarations made that it becomes possible to minimize the chances of running too far afield on these matters.

In an age when information comes fast and furious from all directions—much of it conflicting—it is quite easy to summarily latch on to ideas that simply fail to hold up under close scrutiny. In fact, it is possible to get lost among the assorted voices who are each taking issues down different paths—in many cases proposing mutually exclusive ideas. Without a credible process to evaluate these conflicts it will be difficult to discern an ideas overall value, much like traveling without a map. Those who disdain process in favor of a blinding ideology or tradition, or some other framework that fails to take account of a grounding in the fundamental principles of process, do so at some risk for such sentiments may or may not be on the right track, and without a defining process in place there may not be a way in which to identify blind spots. To follow the heart may offer a place of comfort even if not generally providing a basis for measuring the idea and its correspondence to reality.

If the goal is to optimize such correspondence, then critical processes must be utilized when making assessments. Those interested in processing information with credibility really have little choice but to use a systematic approach that incorporates sensate reason where possible—thereby enhancing the prospects of arriving at ideas that bear some relationship to the real world. After all, thinking that is undisciplined to process does not build modern technology; nor does it lead to knowledge or wisdom.

In the last chapter discussion was had as to the *how* of knowledge. One of the points of that discussion was to highlight its procedural nature, but also to differentiate it from opinion. As a part of that discussion it was noted that most forms of knowledge have limits in as much as knowledge is built upon the foundations of subjective thinking that emerges from pre-logical tacit elements—these being the starting beliefs that frame how the world is interpreted. By combining the tacit part of thinking with human sense and reason, it is possible to come within pragmatic range of reality, and this is knowable by the results that are attainable. The alternative is to employ some approach that lacks the ability to produce consistent and reliable results when comparing one method over the other on real world terms.

In a recent *National Geographic* article titled, "The Age of Disbelief," *Washington Post* science writer Joel Achenbach wades into this subject by reflecting on some of the reasons people fail to employ principled and systematic thinking. He suggests that cultural beliefs are largely connected to tribal identity, with beliefs and values based to a significant degree on emotion and the need to fit in with an identified group rather than on substantive realities. He cites Marcia McNutt, a geophysicist who once headed the U.S. Geological Survey and who is now editor of the journal, *Science*, as suggesting that in many ways people never leave high school in terms of emotional makeup. On this she states: "People still have a need to fit in, and that need to fit in is so strong that local values and local opinions are always trumping science. And they will continue to trump science, especially when there is no clear downside to ignoring science." Achenbach then proceeds to note that the days are gone when the academic community and the publications they serve act as the gatekeepers of information. The Internet and cable TV have had a democratizing effect on information, making it possible to live in what is referred to as a "filter bubble" that lets in only the information that is sought.[1]

In this age of disbelief there are a variety of scientific ideas that are frequently called into question by the lay public—including even by some with levels of science training. Such skepticism ranges from the efficacy of vaccinations, climate science, genetically modified foods, and the age of the earth and life on it, among the more prominent.

Because of the gap between scientific knowledge and popular beliefs, it may be worth considering the ways in which assumptions and process (or lack thereof) can disrupt the human enterprise. A part of this whole equations, then, must go to informational screening skills which can in fact be enhanced by considering the antithesis of "knowledge." Such skills can assist by creating an awareness of specific mistakes of reason. By bearing in mind some of these strategic points it is hoped that readers will have an elevated understanding of how the thinking can get derailed, but also provide important tools for assessing issues that are swirling in the public domain. One of the avenues for accomplishing this objective is to know where the *process* can break down, and that is primarily the focus of this chapter.

Postmodernism

As a beginning place for this discussion, it is worth considering the overarching framework of the problem that confronts modern attempts to interact with the world. It involves methods of thinking that are insidious to the pursuit of knowledge, and as a result leads to the prevalence of ideas that lack a grounded connection to reality.

It was observed previously that the pursuit of knowledge includes a tacit element that hovers as foundational to everything else, and so it is therefore important to begin any assessment of ideas by identifying these elements where possible. Initial assumptions are key to where the thinking processes end up. Some of these pre-logical elements will be identified with some specificity in a later chapter, but for now it is simply important to recognize them to often be tied to self-identity in some way, and thereby tethered by strong emotional attachment. This should add an additional level of caution to those genuinely interested in connecting with reality, because emotions can be blinding.

Even though operating assumptions may feel right, one of the big mistakes is over-assuming their correspondence to reality—the risk being that the logical progression of ideas based on those assumptions may give the

---

1. Achenbach, "The Age of Disbelief," 45.

appearance of incontrovertibility when this may not, in fact, be the case. On such occasions it is possible that the conclusions to be drawn would be rationally impossible if a different set of presuppositions were in play. In short, it is the tacit element, the beginning point that is important to everything that follows. Thus, when critical scrutiny is not directed at these foundational ideas, a significant self-correcting mechanism may be lost with the results being that important data gets dismissed simply because it doesn't mesh with the impoverished starting assumptions.

Building on this important foundational element, the thrust of this book will then be to argue that any dogma or belief—even if right—can never exist credibly without momentarily setting aside the "what" of belief by starting first with the "how" elements that pertain to knowing. In other words, such considerations should frame not only the "*What* do we know?", but the more critical question of, "*How* do we know it?"

Such questions are not trite, with some of the brightest minds on earth spending their lives studying the critical role "process" plays in the acquisition of knowledge. It is the process itself that gets us there—the "how" in knowing that is critically important.

Unfortunately, there are many contemporary impulses that push against these *process* considerations. In fact, such superficiality led one of the leading twenty-first century biologists, Edward O. Wilson, to point out that for the postmodernist, reality "is a state constructed by the mind, not perceived by it," and as such is the polar antithesis of Enlightenment thinking.[2]

The content of that statement needs to be pondered and absorbed, for it conveys powerful insight as to the nature of many of the problems that exist between science and religion.

Essentially, his point is that postmodernists are not terribly concerned with *process* and the important mechanics relevant to the acquisition of knowledge. The focus is all-to-often primarily on the *what*—the end conclusion. This is painfully obvious in a lot of media discourse today where ideological advocates spout all kinds of things without offering much in the way of a supporting foundation. Not only has it filtered into political dialogue, it is also the central driver of many sectarian impulses. *How* questions for postmodernists often appear to be frivolous, or even dangerous when sense-based knowledge interferes with an idolatrous commitment to a dogma tradition grounded in a certitude of mind. This having been said, it is argued elsewhere in this book that in the final analysis it is not the "constructs of mind" per se that are the problem, but, rather, constructs that are unconnected and/or are dismissive of the realities of the real world.

2. Wilson. *Consilience*, 40.

In analyzing how some of this develops in practice, it is arguable that many critical ideas form from within the framework of the tacit, that is to say, a tradition—with tradition then acting as the gatekeeper. To the extent that this happens, the first level of analysis for a new idea goes to the question of whether it is compatible with the traditional understandings. In this way tradition often becomes an obstacle in the way of serious analysis, yet if tradition is to be credible then it cannot just be the gatekeeper—it must also be a part of the analysis.

It is in such innocent seeming ways that ideas get defended against data to the contrary, allowing postmodernism to become the contemporary version of superstitious thinking. In a postmodern world, all that matters are *beliefs*, where the mind settles on certain conclusions—either of the political or sacred variety, or some other category—and data passes through the filter of embraced traditions, supported further by groupthink.

The human urge to seek the refuge of some defining narrative that conveys meaning takes center stage for post-moderns in *belief* itself. Belief is easy for those who put their mind to it, and significantly it requires no necessary correspondence to reality. For those so disposed, process is not essential to belief and can become a seductive shortcut in interacting with the world. Such thinking only becomes evident as inadequate when confronted with questions of credibility and relevance to the factual realities.

In a world where competing values and worldviews are in play, credibility and relevance are important, with a method of systematic evaluation becoming critical. This, in fact, is the role that *process* plays, and on those occasions where ideas lack credible support, they simply become irrelevant to most thinking people.

While many have surrendered to the draw of postmodernist thinking, there are still those who count *process* as both important and strategic to knowledge. With this in mind, the next sections will consider a variety of ways in which the correspondence of mind to reality gets diverted. To understand some of these procedural issues can go a long way towards moving the process in a more credible direction.

Foundation of thinking Blunders

Facts vs. Factoids

Even when there is a general recognition that starting assumptions and process are important components of knowledge, there is another element that can create problems. In this regard, one of the critical areas has to do with understanding the importance of *facts*, differentiating it from non-facts. Very often information gets put forward that is not of a factual nature, even though it may be either represented as factual or is assumed to be factual because it seems to be a part of popular understandings.

Consider one of the paradoxes of modern existence, namely the breathtaking advances in knowledge that have very practical applications to the quality and quantity of life set in the midst of a rising tide of disbelief in many areas of emergent scientific understanding. Most readers will be familiar with a type of thinking that appears to mingle nonsensical ideas into a larger tapestry of understanding that in most ways seems relatively connected to reality. These may be ideas containing a thread of truth, and thereby seem plausible, but which ultimately are not very helpful. As a practical matter many people do not give rigor to cognitive activities and thereby allow credulity to rule some aspects of their lives.

Damian Thompson, in his book titled *Counterknowledge* details a number of ways in which gullibility surfaces. He notes, for example, the success of alternative medicine in a variety of areas that have very little empirical credibility. Thompson mentions, among other things: craniosacral therapy, homeopathy, and reflexology. He also mentions the chiropractic profession where the historic operating premise has been that vertebral subluxation is a significant factor in biological pathology—an idea that is contested.[3] Thompson points out that so-called counter-knowledge spills out in other ways in the form of pseudo-history—literary efforts that masquerade as history, yet fundamentally amount to works of fiction. Along these lines are such literary efforts as *Da Vinci Code*, and *The Holy Blood and the Holy Grail*.

In the age of mass media, facts are often very fungible, frequently getting spun in ways that seem to be responding to preconceptions that offer up the veneer of plausibility, but which in fact are off base in some way. When something resembles a fact, perhaps unverified data, or even invented data that is believed credible—such information is referred to as a factoid, as distinguished from actual facts. For many, facts and factoids sort of mingle together, but for anyone intent upon achieving an affinity with reality, only facts will do.

Perhaps one of the most transparent contemporary examples of a factoid reality, are those who profess disbelief in scientific data that points towards a warming global. From personal acquaintance with folks in this category, it is more of a political posture than an intimate acquaintance with

---

3. The more recent trend within the chiropractic profession has been to move away from the more extreme claims made in this area.

the data. In fact, for many, it does not seem that data is all that important to the conclusions. It doesn't seem to matter that polar ice melts are occurring; that mean global temperatures have been consistently rising; that science has accumulated a lot of data to show that carbon dioxide in the atmosphere does influence climatic conditions, nor that an extensive amount of other detailed data from a variety of disciplines all point to the same conclusion—namely that the world is warming and that human activity is very likely a contributing factor. Particularly important is the fact that once having reached the important threshold of denial, little consideration is focused on the potential consequences—first denial, then end of analysis.

Surely the mature and responsible attitude for everyone outside the field of climatology should be to give some respectful regard to the professional understandings of the vast majority of climatologists and their interpretation of the data—that is to say a consensus of the peer reviewed community of professionals—and then to draw conclusions from all these factors. The more cautious conclusion would be a simple agnosticism on the matter, but even here it is possible to anticipate greater respect of the potential consequences, assuming science to be on the right track. Nevertheless, all too often laypeople are inclined to reach a definitive personal conclusion heavily influenced, it seems, by the vapid emanations of the merchants of misinformation, ultimately sourced by those with a vested interest in the status quo—primarily the oil and gas industries.

Religion is another area where counter-knowledge often seems to find a comfortable refuge—a place where data accessible to sense and reason is sometimes denied in favor of an understanding sourced in sacred writings—where contemporary understandings are assumed to be beyond critique. There are numerous examples, but one present illustration in this regard is fundamentalist Christianity's response to the data on the age of the earth.

On all such issues, skepticism per se is not the problem. Science is built on skepticism, and even invites it. The main issue is that of process malfeasance, where there is a seeming refusal to either engage the data, or to take it and the peer review consensus seriously.

By using the DIKW Pyramid discussed in the last chapter, Jay Bernstein has attempted to develop what he specifies to be the antithesis of knowledge—and in this he would include ignorance, stupidity, folly, misinformation, and disinformation.[4] Each of these terms can have slight variations of meaning, but as Bernstein points out they all tend to represent a category that counters recognized concepts of knowledge. He believes this

---

4. Bernstein, The Data-Information-Knowledge-Wisdom Hierarchy and its Antithesis, 69.

to be an important area of study because non-knowledge is a part of culture and traditional folklore, and tests the boundaries of the concept of knowledge.[5] In identifying the possible inverse of DIKW, Bernstein looks around for an analogous sequence of categories that can be attached to the category of ignorance—a category he terms non-knowledge.

Possible Categories of Non-Knowledge	
Data	Absence or want of data: missing data
Information	Misinformation; disinformation; error
Knowledge	Ignorance
Wisdom	Folly; stupidity

To provide readers with issues that can arise with regards to the definitions that attach to non-knowledge, some of the categories are fairly straightforward, while others are not. He points out, for example, that the term "stupidity" can be controversial with some equating it with low IQ. Bernstein is not particularly partial to this definition and is quick to quote experts who equate stupidity not so much with IQ, but with those who compound ignorance, one on top of another to some ultimate personal detriment. He quotes James F. Welles, for example that states the following with respect to stupidity:

> It is a "maladaptive response to change. In stupidity the response to changing conditions is either insufficient, due most likely to self-deception and the tendency to stick to known ways of thought behavior, or an overly drastic and radical response that is not informed by data. The failure to recognize change that requires response arises from the tendency to insulate oneself from information about changes, which could help one devise an appropriate adaptive response."[6]

Bernstein goes into some detail on each of the other identified categories and the correlating terms, and then ultimately rejects his own hypothesis, that being the idea that the mirror opposite of DIKW has a direct corollary. Yet, he nevertheless believes that there is a category of non-information that can be useful to consider, that can help inform an understanding of the whole DIKW model.

One of the classic books on this general topic was authored by Charles Mackay, titled *Extraordinary Popular Delusions and the Madness of Crowds*,

---

5. Ibid., 70.
6. See generally Welles, *Understanding Stupidity*, (1986).

first published in 1841 and widely reprinted. A more recent work—noted above—has been taken up by Damian Thompson in *Counterknowledge* (2008), where consideration is given to a range of categories—things like alternative medicine, get rich schemes, alternative histories, conspiracy theories, and some of the mistakes of religious thought, such as anti-scientific pronouncements based on a narrow reading of the Book of Genesis. Also, helpful on this topic is the work of Thomas Gilovich, *How We Know What Isn't So,* which makes a number of observations that can guide everyday thinking and discourse by demonstrating specific ways in which people engage in faulty thinking.[7] These mistakes of thinking occur on a range of subjects, including matters related to partisan politics, science and religion, among other. In fact, some of these issues tend to be emotion-laden, leading some to overlook sound reasoning in order to hold on to a comfortable idea.

As has be suggested up to this point, pseudo-knowledge can begin in a number of ways, perhaps as an erroneous tacit assumption, through confusing facts with non-facts, or through errors in process. Errors of this latter category can involve volition and motive—ranging from simple self-deception to outright purveying of deception. Yet, assuming that most people really do want to achieve an affinity with reality, I would like to outline three common mistakes of thinking that Gilovich has identified that can be made by people of goodwill. By steering clear of these common mistakes, it can assist in avoiding dogmatic and erroneous thinking. These points would include the following:

The Mistake of Making Something Out of Nothing

First, humans are predisposed to see order, patterns, and meaning in the world, and this can sometimes drive conclusions in the wrong direction. Such occurrences can be part of a legitimate inquiry involving inferences (something that is a part of all scientific inquiry). But in an effort to support predispositions that are not always on the right track it is easy to latch on to random data and build interpretations that ultimately fail to accurately reflect reality. As beings that seek an explanation of the seemingly inexplicable, it is quite easy to simply see patterns that in reality do not exist. As I was writing this, the mysterious disappearance of Malaysian Air flight 370 was abuzz in the media. This was the plane that was destined for Beijing China, but apparently diverted from its original path and may have ended up in the southern Indian Ocean. People were interested in this story, not only because of the human drama involved, but also simply because it was

---

7. See generally, Gilovich, *How We Know What Isn't So.*

such a mysterious out-of-the-ordinary occurrence. Early on, there were a great many theories floating around out there about what happened, but mostly they were proffered with little evidence—but perhaps just enough to offer veneer of plausibility. Such explanations fit "making something out of nothing"—or very little.[8]

Sometimes conclusions are based on a convenient fit with a given master narrative. It is easy to see how such events can lead astray—given the recognition that a narrative or story plays in the head of every individual that interprets all that happens. When an individual witnesses or experiences something they tell a story in their head about the *what*-and-*why* of the event based on how it integrates into their personal master narrative—regardless of the reality. The first error that might be committed would be failing to recognize that it would not be unusual to mistake such conclusions for being affiliated with "the reality." But by keeping in mind the subjective/objective distinctions a person can sometimes be saved from making critical mistakes in the thinking process.

With this in mind, consider the example of a master narrative that would attribute a weather event to the "hand of God." It may provide a degree of satisfaction to the holder, but it doesn't come with any real-world data to back up such a conclusion. Furthermore, there may well be other explanations that are not antagonistic towards theism, relies on actual data and resists reaching for an indefensible conclusion. In cases where physical data is slim to nonexistent and where the conclusions are under the influence of a master narrative, it becomes important to focus some attention on the master narrative itself. Any failure to do so must be counted as a procedural mistake.

In some cases, where an unsupported claim is made, "motive" should probably be given consideration. Often times the "something out of nothing" argument may be an innocent affair, where the holder does not realize the procedural fallacy of the thinking process. However, there is also a darker side where the argument is based essentially on purposeful dishonesty, where in many cases the argument may rely on a piece of data that superficially appear to be meritorious, but isn't. This data is then spun as a key piece of information, but on closer analysis it becomes evident that the data is not really meritorious at all. Perhaps one of the more prevalent examples occurs with political discourse (both left and right) where assertions are made, backed up by little or no data, with conclusions spun out in a very distorted and dishonest ways—becoming a form of counter-knowledge.

---

8. In later chapters consideration will be given to specific examples within Judeo/Christian theology.

True believers are usually blind to what is taking place. In the same way, dishonest or deluded motives can occur in the religious realm, where people may be quite blind to the reality because a false narrative does not allow the holder to see the reality.

The Mistake of Making Too Much from Too Little

This situation is not completely unrelated to the above prior category, and in fact it is part of a continuum. Unlike the previous example where conclusions are drawn from essentially no data, this class of thinking involves generalizations, where conclusions are vulnerable given that the data set is not complete. While broad judgments made from very narrow specific data can be useful at times, pointing a person in the right direction, they can also lead astray because very often a small data set does not represent an appropriate portrait of a larger more inclusive mosaic. So when working with generalizations from which conclusions form, it is prudent to maintain awareness that such inferences must remain very tentative. For the same reason when others are floating generalizations—perhaps based on anecdotal evidence—it is time for the antenna to go up.

As a category where misadventure can take place, it can sometimes, though not always, goes to improper motives where the intent is to misguide or deceive—again deviating into counter-knowledge. These can be cases where cherry-picked facts provide the veneer of a compelling argument, yet fail in the face of a broader spectrum of evidence. In some cases, a hypothesis may already be in play, and the event in question becomes merely confirmatory. The main problem in this case is that conclusions are being reached without benefit of more comprehensive and representative data that would assure a more objective analysis.

The Mistake of a Person Seeing What They Expect to See

Often times there is a tendency to give a biased evaluation to ambiguous and inconsistent data, though not all bias is bad, and certainly the tacit element always hovers in the background. Honing the skill of discernment between appropriate and inappropriate bias must remain at the forefront of thinking for those intent upon coming to a close proximity to reality. In all of this, there are motivational factors that may contribute to reaching inappropriate conclusions. In general people tend to create their own realities by seeing what they want to see, and within limits believing what they want to believe. Gilovich points out that what a person believes is heavily influenced by what they think others believe, but he also notes that beliefs, attitudes, and predispositions tend to color the estimate of how widely such views and habits are shared by others. Perhaps the classic example on this latter point would be the 2012 Presidential Election where it seemed that pundits and pollsters lived in parallel universes, with most polling suggesting the President's reelection, while a handful of pollsters along with a bevy of pundits—in talk radio land, and on at least one cable news channel—argued that the conventional wisdom was radically wrong. As it turned out, it was a classic case of "wish" influencing "conclusion" rather than actual data. In general, people operate within the framework of a master narrative, and then tend to solicit evidence that supports it, discounting evidence that may run contrary. This is a normal human tendency, and sometimes it can be helpful to simply step back and consider whether or not the master narrative is in need of modification. More will be said about this in a later chapter.

When Agendas Rule

Sometimes erroneous ideas are not the results of some mistake of fact or reason, but are part of a deliberate attempt to undercut inconvenient evidence that would otherwise disrupt a narrative. A prominent part of all agenda driven narratives are religious and political presuppositions. Both tend to provide the framing for how to sees the world, and once these views are set it is all too easy to simply ply the mistakes of reason outlined above as a defense. Along the way it often becomes necessary to just argue away the factual realities that otherwise should mean something.

Having spent my life around the religious community, this chapter is perhaps as relevant as any in addressing one of the prevailing deficiencies of human existence. This deficiency is certainly not limited to the religious community as has been noted above, but common blunders of logic are especially prevalent among those who "know" what they believe, with no amount of data being able to dislodge a given idea. One of the more obvious examples would be interpretation of the early chapters of Genesis, where multiple streams of data seem to matter for nothing—in some cases not even enough to keep the thinking tentative. Obviously, the opposite mistake would be to suggest that science has nailed it down so tightly that there is no possibility that the scientific narrative will modify appreciably in the future. But given the procedural discipline that goes into scientific thinking, there would seem to be an appropriate need for humility in giving it full respect, living with the recognition that the last scientific chapter is yet to be written.

If a scientific idea does not match up with a given religious narrative, then perhaps the fitting attitude is to keep the thinking tentative, with the hope that over time new data will be able to bring about reconciliation. Religious dogma deserves being probed to consider whether or not there might be ways to modify the thinking so as to align with the known data. After all, there is a way of doing this without completely undercutting a faith tradition. Meanwhile, the merchants of misinformation tend to plow ahead in spite of the data.

Certainly, for anyone who is genuinely interested in *truth* there will be little redeeming value in cherry-picking the physical data in an effort to support a particular theological perspective, or even worse, to completely ignore data by claiming that revelation trumps it. History should have taught these lessons a long time ago, yet such attitudes continue to manifest in some quarters.

Finally, for those who have a well-defined conception of God, it can sometimes be easy to rigidly assume that the mind has achieved a certain concordance with the reality even when sensory input is lacking. This is an all-to-human approach, and in most cases should probably be seen as a coping mechanism for humans who seek refuge from the insecurities of life. It is possible to endurance a lot of pain and heartache, and to even put life on the line when the mind is certain about how God interacts with the world. Even where an individual is hitting in the ballpark of reality, there are always reasons for modesty in what is truly knowable on ethereal matters.

Summary Points

Errors can occur by beginning with faulty initial assumptions. The wisdom of sense and reason can help navigation some of this. For example, if the beginning assumption is that a tragedy created by the natural order is in some way a judgment of God, then it will be easy to miss the possibility that it is part of a natural process having predictive qualities.

1. Error can occur in the process itself, and this can be either of a substantive or procedural nature. But the substance of an idea is generally sourced procedurally. Some of the common errors would include the following:
    a. Errors in thinking occurs when concluding on no data at all. Generally, this comes out of the presuppositions that are held;
    b. Error can occur where people are working with actual data. Sometimes it is simply people making too much out of too little

data. Broadly speaking, these are overgeneralizations, and while they are not always incorrect, it is important to be aware of what is being done, and adjust the level of certitude that is attributed to it. Generalizations can be appropriate when done in context, for they in fact can serve as useful shorthand when a broader array of data is not available.

    c. Error can occur for people who tend to see what they expect to see, giving very little regard to data that does not fit the operating narrative. In this way, belief supersedes the reality.

2. Agendas can drive faulty thinking, and have important applications to ideas that arise from religious dogma.

CHAPTER 3

Science as a Candle in the Dark

Chapter Summary

Whenever there is a conflict between a sacred belief and sense-based data, it is not uncommon for those in the faith community to assume that there is simply something wrong with the scientific understanding—this being the conclusion of any further analysis. Sometimes a conclusion of this sort is arrived at by those without a substantial grounding in the scientific method. In this chapter, the objective is to lay out the methods of science, and do so in a way that is accessible to the lay reader so that there will be less of an inclination to summarily dismiss science. The process of discovery that humans engage in has been extremely successful at fulfilling the human longing to understand the mechanical features of the universe, including systematic processes that can be tracked back to the remote past. But science can answer only the "what is" questions, and comes up short on "ought," "purpose," and "meaning" questions. To understand this differentiation can go a long way in reducing tensions between science and religion.

There is a bit of hubris associated with a non-scientist attempting to write about the scientific method, yet that is what I propose to do in this chapter. In my defense, while my formal education was not in the sciences, I have spent quite a number of years now giving independent study to the methods of science, and consequently do have some strategic understanding of it. In addition, this manuscript has been reviewed by a couple of academic scientists, and they have been helpful in assisting me to add sharpness and nuance to the discussion.

For me, this quest has been personal in as much as my faith tradition, discussed earlier, includes certain ideas that seem to run contrary to physical data. So this mission has been an attempt to sort out the reality, and to search for ways that a faith tradition might be reconciled with conclusions of science. I started this quest with a bias towards the understandings of my faith tradition. But I was not merely on a quest to engage in self-deception by cherry picking the data that supported my religious presuppositions. Thus, a critical part of this entire process was to consider the methodology of science so that I would be in a position to fully understand the significance of scientific conclusions that differed from my faith tradition. This process necessitated that I consider all the data that might inform me—both supportive and non-supportive—about my inherited biases.

In coming to this particular conversation, I am mindful that some expect a "sacred tradition" to shape their entire worldview without any intervening mediation. In such cases no amount of data will ever change a mindset of this order. Perhaps the following bumper sticker sort of frames this approach, "God said it, I believe it, and that settles it." Such attitudes no doubt speak to the human hunger for certainty, and it is important to recognize that mental certitudes of this sort perhaps do serve the purpose of reducing the anxiety of uncertainty. Clearly such thinking pervades parts of many religious communities, and this becomes particularly evident when scientific findings do not line up with a treasured interpretation of a sacred text such as Scripture. The battle lines get drawn, and war is sometimes declared on the science community.

Those who do not have a basic understanding of the methods by which scientific knowledge is acquired are often quite ready to sacrifice it, perceiving it as being in the way. As noted at several earlier points, I have firsthand knowledge of a spectrum of Christianity that interprets the beginning chapters of Genesis in ways that completely ignores a vast amount of data that should inform any reading of these ancient texts. It seems that the opportunity for confusion on some of these issues can only be compounded for those who do not have a basic grounding in the scientific method.

In the midst of such conflicts, there may be some value in stepping back to consider some of these matters in a more strategic fashion. In this regard it can be helpful to examine the progression of human civilization, it having emerged from a long dark past of superstition, where the forces of the natural order was deemed to be the work of the gods. In antiquity when disaster struck, it was obvious that God, or the gods, were angry. Alternatively, when events of the natural order turned in favor of life, divinity was obviously demonstrating pleasure. Over time these primitive ideas have largely been displaced through discovery that the universe is governed by the regularities of cause and effect, and that as such, logic and reason applied to sense-based observation can go a long way in figuring out the natural workings of things.

The emergence of a methodical process by which an understanding of the world could be acquired generally came with recognition that while there are a few certain absolutes, most of what is knowable is achieved through connecting the senses to the material world, and doing so in a rational and systematic way. This is, in fact, the foundation of scientific discovery, which has enjoyed phenomenal success in unlocking many of the mysteries of the universe. The unparalleled successes of this enterprise validates the overall importance of its methods, exposing a certain level of folly for those who choose to run against it, and this particularly applies to non-experts who opine about things of which they know little to nothing.

Those who recognize the power and value of science may consequently sometimes find themselves in a quandary over seeming conflicts between scientific findings and religious belief. For sure these can be difficult occasions, and those schooled in the scientific method may be better equipped to make necessary adjustments when the evidence compels. But for those who rely exclusively on religious dogma, old ideas are often hard to shake even though conflicts may be resolvable by simply having more flexibility in the interpretation of a sacred text.

As consideration turns to the whole question of "knowing," those who value religious faith may sometimes feel the sense of not being respected when it is discovered that under most epistemological formulations, religious belief does not fair very well on the traditional ladder of knowledge. This has nothing to do with bias or animus per se, but in the nature and source of religious belief itself. This point will be discussed in more detail in succeeding chapters, but in order to build an adequate foundation for later conversation it seems at this juncture the most helpful way to proceed is to consider exactly why science is so successful. Once this is understood, it will be easier to comprehend what humans are up against in conveying gravitas to the realm of religious ideas that run counter to the interpretations of data.

A number of religious issues overlap with science that could be considered, but one in particular that was mentioned above involves the question of just how long the universe, earth and life have been around. Historically much of the Judeo-Christian community held to the view that the Genesis creation story was an event that occurred around 6,000 years ago. Though Scripture itself does not date this event, there have been several attempts to put a date on the "creation event" by use of Biblical chronologies that provide details regarding the ages of the patriarchs listed and their descendants. Perhaps the best known of these chronologies comes from Archbishop James Ussher from the seventeenth century who developed what became one of the most widely circulated creation-dating attempt, finding its way into most of the King James Bibles through the end of the eighteenth century. It was his conclusion that the Genesis Creation occurred in 4004 BCE, and consequently became a widely shared approach within the faith community well into the nineteenth century.[1] Furthermore, it is noteworthy that even the scientific establishment through most of the eighteenth century was generally comfortable with notions of a relatively recent beginning since there was not an overwhelming amount of known data at the time that pointed strongly in a different direction. It was only as the evidence accrued for the antiquity of the cosmos, as well as earth and life on it, that this earlier idea became dislodged. Thus, this earlier view within the sciences has fundamentally changed, and it changed because of a consistent and overwhelming amount of data that has been recovered over time that strongly suggests far different conclusion. Consider some of the following ways in which scientific dating occurs:

- **Light speed**—because of the vast distances of cosmological space, distance is measured in light-years, which is simply the measure of how far light travels in the space of a year. It is interesting to note that light can be a useful measure of time for observations at great distances, for we can observe stars out to billions of light-years distant. What this means is that these observed stars are not now as we see them, but are how they were billions of years ago. In fact some of these stars at great distance may have vanished in a supernova death spiral millions of years ago, but we don't know it yet because the light of the explosion event has not reach us. This measure then becomes prima facie evidence that the universe was created a very long time ago for if these far away stars had been created a mere few thousand years ago, they might be there, but we simply would not know it because they are too far way for the light to have reached us.

- **Radiometric dating**—all matter, from organic to the inorganic, is composed of the elements of the earth. All elements have an atomic weight based on number of neutrons, protons, and electrons that compose the nucleus. Some elements of the same type may have a slightly different weighs due to the number of neutrons in the nucleus, and some of these configurations are unstable resulting in radioactive

---

1. See, for example, http://en.wikipedia.org/wiki/Ussher_chronology; see also generally Davis A. Young & Ralph F. Stearley, "The Bible, Rocks and Time."

decay into more stable forms. There are actually several ways in which this can occur, but the details are outside the scope of this chapter. The important point to grasp is that the rate of decay of these unstable elements all fall on a reliable exponential curve and consequently can be used as a clock by which to measure time. Some individuals attempt to cast doubt on the reliability of this measuring method because, for example, how can it possibly be known that a radiometric clock is still keeping time at the same rate as say a few million years ago? Well, the short answer is that each element has a different half-life (which is the time it takes for half of the unstable elements to decay), some having very short time measures of a few weeks or months, and some out to millions of years. But every element follows the same exponential decay curve that can be plotted, thus proving to be a reliable measure of time. In fact it is common practice to use more than one element as a method for cross checking a time measurement.

- **Ice cores**—sometime back it was discovered that glacier ice cores provide a measure of time, with them providing a visual annual banding due to seasonal variations in weather that offer a dateable history out to several hundred thousand years. As the cores get deeper, there is compaction that takes place, resulting in the rings get increasingly thinner and closer together. Consequently, by the time the core gets out to around 100,000 years, the dating error rate goes up. Some of the dating tools used in the process include: 1) cross-checking the analysis of the chemistry within the core itself. This would include analysis of air temperature variations that will be evidenced by the molecular weight of the water molecules that vary with the temperature, thus providing some independent seasonal indications. The way this works is that the water molecule can have slightly different configurations with heavy water containing a heavier hydrogen isotope, resulting in water around 11 percent heavier than normal water. Because of this heavier composition, and the fact that moisture coalesces at different rates depending upon temperature, the amount of heavy water present can provide a measure of the temperature at the time the moisture fell; 2) dust and pollens in the air tend to be seasonal, again helping to validate annual layers; 3) dating can be authenticated by known historic events, such as major volcanic eruptions. In fact, such events themselves have signature isotopic configurations that can help tie them to the event itself; 4) specific isotopic analysis of core samples; 5) finally there are other tools used but the ones mentioned above would be the most common methods.

- **Seismic data**—Alfred Wegener, a German meteorologist and geophysicist proposed in 1912 the idea of continental drift. This idea came from the observation that the shapes of the American and Euro/African continents were such that they could roughly fit together sort of like a jigsaw puzzle, and this is particularly true if the continental shelf is taken into account. Furthermore, the flora and fauna associated with the comparable contact points of each continent are very similar. Thus was born the idea of continental drift. The science community did not warmly receive this idea at the time it was first floated, given that there was no known mechanism by which this could be taking place. However, it has since been discovered that in fact the continents are drifting due to seismic activity, and that the European and American continents are moving away from each other at the rate of a few centimeters a year. On the assumption that one super-continent existed at a remote point in time, and assuming the current rate of movement—this has been a process occurring over the period of millions of years. Even if one assumes that continental drift may have been much faster at some earlier point in history—assuming such in an effort to hold to a young earth—there are so many other evidences of an ancient earth that such efforts will be of little value.

- **Ocean sedimentation**—the oceans teem with microscopic organisms that go through a life cycle and then end up as sediments on the ocean floor. Among these are coccolithophores and foraminifers, and both possessing a large amount of calcium carbonate that consolidates and eventually hardens on the bottom of the ocean floor over a period of time. The size of these organisms is so small that it takes a very long time for the build up to occur. In addition to the observation of the ongoing rate, other dating methods can assist in developing timelines for this process, which point to time extending out to 10,000 years when coring into the sediment on the ocean floor by about a meter. This must be put up against the fact that the sediment is a kilometer thick in places, leading to the compelling conclusion of the great antiquity of sea life.

- **Coral reef growth**—based on annual growth rates and other dating methods, some reefs, which grow based on the life and death cycle of living organism, can be dated out to hundreds of thousands of years.

- **Geologic column**—the geologic column has been found to have a general order running from very simple life to increasingly complex life at the higher end of the column. For sure, there are unconformities that some will argue demonstrates the bogus nature of geological science, yet there is generally a good understanding as to the type of events that result in such occurrences. As geologists are quick to point out, the order found in the geological column is used commercially every day in the search for natural resources, including oil, and natural gas, etc. In fact, the antiquity of the fossils was established well before radiometric dating was discovered, but its discovery did add a confirmatory element to these prior conclusions.

There are other dating methods, such as dendrochronology and racemic dating, but the ones sketched out above represent some of the most prominent methods that have led the scientific community to the conclusion that the universe, the earth, and life on earth have been around for a very long time. Because of the power of science to develop tangible applications of knowledge, its significance is attached to this discussion so as to impress upon those unfamiliar with the scientific method, exactly how science goes about acquiring knowledge. In this regard it will be a somewhat didactic chapter, and those who are competent in this area may wish to skip on to the next chapter. However, there will be a few key points addressed that will be helpful in constructing a credible worldview. So with this introduction, perhaps it is now time to turn to the methodology that has made science so successful.

The Structure of the Scientific Process

A couple of chapters back it was noted that the type of knowledge science brings is usually considered reliable due to the processes employed, though this can never guarantee against the possibility that a broader, more inclusive explanation will surface to displace previous understandings. As Thomas Kuhn pointed out, paradigm shifts do take place when sufficient data is obtained to push scientific conclusions in a modified direction. In fact, this seems to be one of the key initial distinctions between science and the faith community. Science has a nimbler attitude towards "truth" that promotes a shift in understanding as the data demands. Religious truth, on the other hand, is sometimes viewed as timeless and unalterable, even though it is permeated by interpretations and theories; and even though new information and insights may suggest the wisdom for more flexibility—including revisions of understandings. It all boils down to whether human perception of reality—scientific or sacred—is to be understood as dynamic or static. If it is thought of as static, it will be seen as unalterable. For science, and for epistemology generally, such matters are always deemed as dynamic in nature. Understandings for finite humans are never timeless or unalterable.

In the example above, some detail has been proffered regarding some of the data elements that show why a 6,000-10,000 year timeframe for the emergence of the universe and earth are not scientifically viable. The purpose was to provide a high level overview of why science has settled on a very different idea from that held by some Christian traditions. As this chapter delves into the processes of the scientific method itself, the intent is to develop some of the important details of the processes by which scientific ideas emerge. Such clarity is worthwhile, particularly amidst conflicting religious constructs, in as much as offers up some basis for assessing not only the validity of the scientific method but also possible ways conflict can be ameliorated.

In spite of the fact that scientific knowledge is never absolute, as previously discussed, it has had phenomenal success in providing a world unthinkable a century ago, and for this reason cannot be dismissed. Science and its applications are now integrated into virtually every aspect of modern life—including computers, smart phones, air travel, and high tech health care—just to name a few. Contemporary life is truly remarkable when contextualized by historic standards. This reality should be a cautionary point for all sectarian inclinations to swim against the currents of science on the idolatrous presumption that a given interpretation of revelation is somehow infallible and therefore unassailable.

While the results of science are everywhere, the public understanding of the practice of science is not so common. For this reason, I would like to briefly outline sort of a high-level description of its methods, as I understand it, including the activities and modes of thinking common to scientists in all disciplines. Readers not already familiar with the scientific method should review this and the prior chapters thoroughly before proceeding, since these are foundational to some of the issues that will be discussed later.

Scientific Methodology

Observation

All science begins with an observation of some aspect of the natural world, with this activity occurring by reliance upon the five human senses—seeing, hearing, tasting, touching, and smelling. Its significance to the methods of science is based on its operating premise that the universe operates on the basis of cause and effect. From the process of observing, a question will sometimes form as to causation of some observed effect, and with a little creativity one or more hypotheses emerge that offers to explain the observation. Thereafter, the search is on to test the hypothesis.

Hypothesis

When making an observation of some phenomena, science takes whatever minimal evidence may be available to explain the observation and then uses reason to create a proposed explanation. In physics and other scientific disciplines, the hypothesis often is framed around a causal mechanism or some identified mathematical relationship. One of the primary criteria required for a valid hypothesis is that there be some basis by which it can be tested, and that the test can be replicated. The hypothesis can be generated in a number of ways, but often it is as a result of inductive reasoning that proposes the explanation for the observation of some phenomenon. Sometimes the precursor to a hypothesis will be a "why" question, with the hypothesis attempting to propose an answer.

The commonly held view of the average person is to presuppose that science collects evidence to prove a hypothesis, yet factually a properly constructed research project does the exact opposite—that is, energy is expended to disprove the hypothesis. Science and statistical texts are quite consistent on this approach and should be understood as a very important point. The reason for setting up the test this way is that it helps to keep in check investigator bias. While it is sometime possible to find the process done in reverse order, it is not considered as the normal practice, and will likely receive enhanced scrutiny in peer review. The reason this approach is the norm is that humans are naturally biased to see things that are expected and to not notice conflicting observations. A single exception can disprove an idea that a researcher is attempting to validate. For such reasons it is recognized as good practice to try very hard to unsuccessfully disprove a hypothesis, for if one tries, but fails to disprove it, it only strengthens it.

Thus, a red flag should go up when the proponent of an idea does not discuss problematic data, or ignores evidence that may point in a contrary direction. This might include, for example, a young-earth creationist who mounts up reasons for holding such views, but fails to mention any of its scientific weaknesses. In the main, while there is seldom-absolute perfection for validating scientific ideas, the peer review process can be brutal in ferreting out an ill-conceived research design. The specter of critique becomes a powerful motivator for an investigator in developing competent methodology. When comparing the rigor of the scientific method with the rigor in ordinary discourse, peer review stands up quite well.

Theory

Over time, if there is no success in disproving the hypothesis, it may eventually advance to the next stage and become a theory. The primary trait of a hypothesis is that something can be tested and that those tests can be replicated, where multiple scientists can examine an idea and end up with the same results. This approach helps to ensure the integrity of the experiment. It is a process that can take years, and it is a given that many hypotheses never become theories since it is sometimes difficult to gather sufficient supporting evidence, or the evidence acquired in some way invalidates the idea. But if falsification is unsuccessful after repeated testing, a theory may be born, becoming accepted as a valid explanation of a phenomenon.

Theory is thus, distinguished from hypothesis in that it does come with a fair amount of supporting data, thus, affording some confidence that the investigator is on the right track. Theory is differentiated from data in that data is observable and measurable, while theory is an explanation of the observation backed up by a body of supporting data. Theory can modify over time, but data itself does not, though more sensitive instruments can many times modify a previous interpretation of it.

Laws and Principles

Occasionally, enough evidence accumulates that theory gets elevated to the level of scientific law. Quite often these laws are reducible to mathematical equation, and are universally accepted by the scientific community. Nevertheless, a scientific law never reaches the point where it is above being questioned or challenged, though seldom are refuted.

An outstanding example was Copernicus's supposition that the appearance of the sun in the east, its arc over the sky throughout the day and then its disappearance in the west was perhaps explainable by the rotation of the earth rather than the rotation of the sun around the earth. Another example was Isaac Newton's recognition that a falling apple and the motion of planets have something in common in spite of the vast differences in application, the common denominator being gravity. These examples started as hypotheses that ultimately matured into theory, and then later became settled laws of understanding as the weight of evidence continued to mount.

Distilling from this section, then would be the notion that when an idea has been elevated to a law of nature, it comes with an overwhelming amount of evidence that has thwarted all attempts to disprove it.

One finally point here. Sometimes even the scientific community uses these terms incorrectly, though it is done knowingly and not as the substance of confusion. One vivid case in point involves the physics community's attempts at developing some understanding of the nature of matter at its most fundamental level. One of the ideas is that at the most fundamental subatomic level matter consists of vibrating strings, also known as *string theory*. For starters it technically has not advanced to the level of theory, and at most can only be considered a hypothesis. But the problem here is that string theory appears as an idea that has an incalculable number of possible variations that in most cases are currently beyond testing. As such, even to call it a hypothesis is a bit of a stretch in the classical sense; though over time it is possible that some parts of this idea may become testable, in which case its status could be elevated. There are also problems with the idea that reality may be composed of many universes—sometimes referred to as a "multiverse hypothesis,"—in this case there really is no way to test this idea, since it seems likely that the only universe humans will ever have the capacity to know is the one they reside in. There is simply no conceivable empirical data by which this idea could ever be tested, so technically it is not a scientific hypothesis, nor is it science—mere speculation about how physical reality may be constructed. Similarly, it is for the very same reason that the "God hypothesis" can never be elevated to the level of science, and this irrespective of predilection in this regard.

The Observational Basis of Science

The powers of the five human senses are often taken for granted. Humans see, hear, smell, taste and touch, yet generally do so without any thought of what life would be like in their absence. To contemplate for a moment, the existential condition of a person who did in fact lose these faculties, it is possible to gain some insight into the power they represent. Without such sensory inputs, a human becomes nothing more than an isolated mind, with no ability to interact with the outside world—no means of any two-way communication. In such a situation, the person would be locked in the ultimate prison, and for sure any ability to learn anything about the world would be dead. So sensory input matters very much to the human ability to learn about the world. When the senses are combined with cognitive capacities for reason, science becomes possible.

Observations of the world allows for learning about how it functions, and are only possible through the use of the five senses. When observations are made and recorded, this information is generally referred to as "data."

Early on, the only observations possible were by direct means without any mediating instrumentation. It wasn't really until the discovery of optics that it was possible for great strides to be made. Although the ancient Egyptians and Assyrians, and a few others were working with optical lenses around 700 BCE, it wasn't until the twelfth and thirteenth century in medieval Europe that the scientific value of optics really began taking off. With optics humans for the first time discovered a world of the very small—things like bacteria and viruses, and ultimately with the most powerful of optics—atoms. Optics also afforded the opportunity to see into the far reaches of the universe. It was Galileo, in fact, who was able to confirm the validity of the Copernican idea, proposing, contra to the then prevailing scientific belief (as well as the interpretation of a number of Scriptural passages), that the earth was not the center of the universe, nor was it fixed on a firm foundation. It was Galileo's discovery of four moons circling Jupiter that was a game changer of gigantic proportions, allowing for a paradigm shift. In the modern era, the human capacity for observation has been extended even further with sophisticated electronics for measuring chemicals, the strength of radiation, and other environmental phenomenon that would otherwise be undetectable, as well as computers to record a massive amount of data with accuracy.

Science is a Social Process

Perhaps one of the more important ideas related to the scientific method is its social structure. There is an entire process in place that is used by which ideas achieve scientific currency. It is social in nature and considered to be one of its most important features—due to the possibility of bias. Because of the subjectivity of the senses, bias is a normal part of what it means to be human, and consequently it is easy to overlook critical elements that could affect conclusions regarding a given hypothesis or theory. So scientific ideas get published and critiqued by a group of peers. The best ideas survive in-toto, or survive with modification. Others go down to defeat. As a social enterprise, science has built in a self-correcting mechanism that is useful in overcoming natural bias. Of course, in the real world this is never entirely possible, with the best that can be accomplished being an awareness of the problem and then to develop methods to help control for it. On this point, a physicist friend of mine, Mailen Kootsey puts it this way: " . . . the social self-correction mechanism combined with the ultimate tie of theories back to real world data keep errors in scientific thinking to a minimum."

This process also provides lay people with some way of assessing ideas that surface periodically. For example, it is not uncommon to hear someone promote a far-out idea that may refer to "a science expert" as the source. Two important observations should be made on this. First, sometimes in the mind of the average person, any individual with science credentials in any given field is viewed as an expert qualified to opine on all scientific issues. So the first red flag should go to consideration of the credentials of the individual who is floating a seemingly unorthodox idea. The question that must be addressed is whether the originator of the far-out idea being offered is a subject-matter expert on that particular issue? If not, then there is generally little need to go any further. If the person in question is a subject-matter expert, then it is appropriate to move on to the secondary consideration, namely whether the unconventional idea has been submitted to peer review. The point is, while more credibility goes to an "expert" opining in his or her field of expertise, this alone is no reason to believe the idea is meritorious. Any "expert" who talks up ideas that run counter to the prevailing view must be regarded with caution. There are a lot of charlatans out there.

Let me mention one popular issue that is a case on point—an issue touched on earlier. The vast majority of climatologists (by some counts 97%) believe that global warming is a reality, and that human activity is contributing to this reality. If this was a medical issue involving the propriety of a certain therapy, there are few patients who would buck these odds. Yet when it comes to global climate change, a great many laypeople seem to think they have greater wisdom than the peer reviewed experts. Of the talking-head skeptics, I have yet to hear an argument that delves into the actual data itself—and the data should be the thing that really counts in such discussions—not the wisdom of the climate expert who may be making a different argument from that of their peers. Unfortunately, this issue has become political and emotion laden, and generally devoid any empirically based discourse.

I personally have little stake in the outcome of this issue other than my desire to see the long-term survival of biological life, but since the survival of biological life is ultimately at stake on this issue, this is a very important topic. As a layperson on climate related issues I simply have no choice but to rely upon the peer review process of the scientific community to be providing a reasonable proximity to the reality. There is certainly nothing wrong in maintaining a skeptical attitude, or to ask probing questions—in fact the scientific method would invite it. However, rank ridicule of climate change science without ever engaging the actual data exposes a level of ignorance within a large swath of the public that is rather shocking. The reality—whether it happens to be climatic warming, cooling or some other—will have the last word, and if the current scientific consensus is correct, history will not be kind to those who have ignored the warnings and neither will the consequences.

Another situation in which the social processes of science are important to laypeople is in assessing the reliability of arguments made by people with science training. There are a range of situations in which the idea from the so-called "expert" should be held with healthy skepticism. These would include some of the following: 1) the expert is pushing an idea that is outside the mainstream of peer review, or 2) the idea has not gone through the peer review process, or 3) the idea has been rejected in peer review, or 4) the expert is on the payroll of an organization that benefits from the contrarian view—all such situations should be deemed as red flags.

This same principle applies to those who assert expertise in a field outside their formal academic training. I am aware, for example, of a physician who has a website that promote a range of fundamentalist creationism ideas, who talks in scientific language giving the veneer of expertise, yet his overriding thesis should be viewed with a degree of skepticism—in as much as many of the ideas he promotes have either been rejected by, or have not gone through the peer review process. The fact that he passes off as an expert on subjects outside his actual area of formal study should be grounds for considerable skepticism.

I have discussed this website with several genuine subject-matter experts who are familiar with it and the general consensus is that on some points its author is hitting in the ballpark, while on other points he is demonstrably off base. So, if the objective is merely to seek confirmation of a bias then websites of this caliber can be a useful resource, but anyone genuinely interested in achieving a close proximity to reality should embrace its ideas with extreme caution.

All of this, of course, raises the question of *process* for those who may believe certain scientific findings to be at variance with an interpretation of Scripture, and who therefore deem the elevation of science as somehow a negative reflection on spirituality—as abdicating a commitment to the sacred. To the extent that such attitudes exist, it may be worth revisiting the epistemological basis for elevating the senses and the scientific methodology in the first place. In short, it has a track record with tangible results. While there is always a possibility that troubling scientific ideas will modify in time, there is also a possibility that human understanding of an inspired text will also modify as new levels of understanding emerge.

Science Proceeds by Experiment

Science relies on the senses to observe and then to reach conclusions about the observations of physical phenomena. This process generally occurs within the framework of testing an idea or hypothesis and then reaching certain conclusions based on the observation of what happens. In any case where such cannot be tested by experiments or observation, the idea is technically considered to be outside the realm of science. Even prominent idea that scientists talk about sometimes fall into this category—ideas such as those mentioned above—namely the proposal that there exists a multiverse, or quantum string theory, for the simple reason that these are ideas that cannot be empirically tested at this time.

The Tentative Nature of Scientific Knowledge

As discussed a couple of chapters back, even after a scientific theory is formulated it is always considered tentative and open to modification in the event that new data emerges, but in general this usually happens with small incremental changes rather than whole-sale modifications. It is on this point that some argue the intrinsic inferiority of science. Yet, this open approach to "truth" has proven itself over time. In commenting on the methodology of science, consider the observations of physicist Mailen Kootsey in his following comment:

> Just like friendships, confidence in a scientific theory grows as time passes with no significant explanatory failures and continued success by the theory in matching quality data. Sometimes, though, a cherished theory fails in the face of new data. A colleague of mine used to say, 'the cruelest murder is the killing of a beautiful theory with a single fact!' There have been many examples of extended arguments in the scientific community about controversial experiments and theories. In September 2011, 171 scientists from an international collaboration announced that thousands of measurements made between 2009 and 2011 seemed to show that neutrinos (exotic, massless, sub-nuclear particles) from the CERN laboratory near Geneva, Switzerland, were reaching a laboratory in central Italy some 60 billionths of a second (60 nanoseconds) earlier than light would. This finding was at odds with Einstein's special theory of relativity, which says that the speed of light is the upper limit attainable by any particle or wave. The first repeat of this experiment seemed to show the same effect. More testing finally revealed that the

anomalous result was actually caused by a loose cable connection—so Einstein's theory of relativity continues to stand. This scientific saga is documented in a series of short articles and notes in the prestigious journal *Science*.[2]

Ideas that come to be accepted as part of the body of scientific knowledge are arrived at based on the best explanation of the data at the time, and represent a consensus of the science community. Anyone is welcome to introduce an alternative theory, but in order to receive attention it would have to be as good or better at explaining the relevant data if it is to be taken seriously.

It does happen that findings from studies sometimes turns out to be flawed in certain particulars, requiring later modifications. This suggests wisdom in having an attitude of tentativeness, even while remaining mindful of the overall credibility of the scientific enterprise given its transforming effect on just about every aspect of human life. As noted by Kootsey, "ideas that come from other non-data based sources—tradition, intuition, personal bias, etc.—ultimately cannot win because science is rooted in actual observations. Science can also tell us something about our past and our future, but these areas are also the purview of philosophy and religion."

The Mechanics of how Scientific Ideas Develop

Early in the history of science, observations of data were unmediated, that is, observations were direct in nature and therefore relied upon no instruments. However, as optics became available progress in scientific knowledge made rapid advances in as much as it enabled humans to see that which the unaided eye could not see. Over the intervening centuries, in addition to optics, many other important instruments have greatly advanced scientific endeavors. In the sections that follow, some of methods that have advanced human understanding will be considered.

---

2. See http://spectrummagazine.org/blog/2012/07/09/bringing-real-world-genesis-what-makes-science-valuable-part-1-observations-experime I have cited Kootsey at several points in this chapter. For one thing, he and I worked together on the *Spectrum Magazine* blog article series, and have respect for his thinking and scholarship. Secondly, he is in good standing with the community of scientists, receiving his terminal degree in physics at Brown University, with academic appointments at Duke University, Andrews University and Loma Linda University. He thus has the ability to speak authoritatively in a way in which non-scientists are unable to do.

Reductionism

The practice of reducing a process of nature down as a method of understanding how it works has been one of the principle means by which science develops knowledge of the system under study. In addition, each component part is studied as an independent unit. For example, in biology processes are separated out on the basis of cells, and molecular processes, down to the lowest identified subcomponents—ultimately to DNA. This process leads to a hierarchy of physical scale and complexity. Kootsey observes on this point the following:

> Scientists have found it necessary to concentrate their work on one level of this hierarchy, considering relationships with one level above and one level below the level of interest. For example, in living systems, cell biologists think about how their cells work together to create a functioning organ (one level higher) and what biochemical systems exist in their cells (one level lower). They do not usually try to understand their cells in terms of sub-nuclear physics or sociology of individuals (both multiple levels away). There is a desire to understand the effects of genes up several levels in the biological hierarchy for the purpose of understanding diseases, but these studies are not yet on solid theoretical ground.[3]

More recently, there has been a growing recognition that not all the explanatory arrows point downward. Perhaps the most significant figure in this area of study is Stuart Kauffman, a theoretical biologist, who argues that synthesis, the process of putting things together to create a more complex whole, can produce an almost infinite number of variations.[4]

Models

Scientific ideas that attempt to explain what is going on in the real world, serve as models. Field experts work with different kinds of models, depending on the amount and quality of observations (data) required. The type of model used is relational to the level of detail needed. Kootsey has identified several types commonly used and they can range from simple to more sophisticated. Among these he would include: a) names for the object or event being studied; b) pictures, drawings, videos and animations used to convey information; c) physical mock ups, and perhaps one of the most famous

---

3. Ibid.
4. See generally, Stuart Kauffman, *Reinventing the Sacred*.

of these was the double helix DNA structure; and d) mathematical and computer simulations where a range of variables can be controlled. Kootsey notes that mathematical models are so significant that they command high respect wherever applied.

Kootsey further identifies several important characteristics of a good scientific model. He notes that they a) "must reproduce the relevant observations, with accuracy at least comparable with the observational error limits; b) they must have predictably emulate the behavior of the system when applied conditions change. Predictions are an important goal of science and it is the model that makes predictions possible. Names, static drawings or pictures, videos, and animations are clearly unable to make predictions for new conditions. Thus the higher forms of models—physical or Mathematical—are necessary to meet this requirement; c) the model must be falsifiable. If there is no way of disproving the model, then it is not useful."[5]

Assessing Alternative Models

When two or more models are proposed there can be ambiguity over choosing the model that fits the observations the best, or in choosing the model that makes the most accurate predictions. In such cases the science community employ the law of parsimony, also known as Occam's razor, named after the fourteenth century English Franciscan friar Father William of Ockham. The rule says simply: if there are multiple explanations or models that work equally well, choose the simplest one, as there is no need to add complexity to a model if the added complexity does not improve the model's performance.

Philosophical and Methodological Naturalism

Brief mention was made earlier to methodological naturalism. However, naturalism is often packaged in two distinct forms—methodological and philosophical. Naturalism itself involves the idea that only natural explanations can be used in scientific models, with the most severe articulation of it being the idea or belief that natural law is the only force that operates in the universe. This view may sometimes be referred to as scientism. Those who reject this approach point out that it takes a certain amount of arrogance to conclude in such a categorical way, particularly in the absence of proof. Methodological naturalism avoids the extreme claims of philosophical

---

5. http://spectrummagazine.org/blog/2012/07/30/bringing-real-world-genesis-what-makes-science-valuable-part-2-developing-scientific.

naturalism by simply holding that whatever science can know, is only available through sense-based strategies that employ reason. It has become the working method of science, and is approached in a way that avoids ultimate philosophical characterizations. On this Kootsey makes the observation that a scientist can believe in God and still follow the rule of never incorporating into an explanation or theory, the supernatural.[6] There is an important reason why this should be the standard for all scientific investigations, and that is that there are repeated examples from history of humans having invoked God as an explanation that turned out to have a very natural justification.

Nevertheless, it is not uncommon for those who come from a religious tradition to refer disparagingly to "secular science," particularly when the science community comes to some understanding that may be at variance with a sectarian belief or interpretation of Scripture. Such term is used as sort of a pejorative—secular science as opposed to divine truth. First, it should be clear from this chapter that all science is secular in that it is sense-based. This, in-and-of-itself is not anti-religion, but simply does not rely upon religion in reaching its conclusions. But there is a very good reason why religious people should not oppose science, for if there is the belief that the universe was created by God, that God is a being of logic and order, and that he created humans with the ability to come to understanding through the use of sense and reason, then God's book of nature will not seem quite so hostile to revered sacred revelation. It will become intuitively understood that the book of nature and the book of revelation are not opposed to each other. Ultimately, the person of faith may discover there to be wisdom in finding a way to understand both books in some unifying context.

Summary Thoughts on Science

In this Chapter some of the specific methods of how science goes about its work have been outlined and even though scientific conclusions are reached that are based on the unavoidable subjectivity of the senses, the process nonetheless involve objective tools and means. When a broad range of trained individuals using instruments to enhance observations and interpret data, with most coming to the same overriding conclusions, the chances that they are on the right track is greatly enhanced. Because the scientific method involves the study of natural processes at work, cause and effect become important factors in both anticipating the future as well as understanding the past in many important respects. The fact that science

6. Ibid.

has had such phenomenal success in this regard is prima facie evidence that the overall methodology is effective.

When a given idea reaches full maturity, backed up by observational data, the conclusions are termed "scientific knowledge." This terminology is not merely empty rhetoric, but represents ideas and concepts with practical significance, given the ability to understand the causal basis of some observed effect empowers accurate predictions. In this way knowledge becomes power, and brings with it strong assurance that humans are on the right track.

The entire process of systematic discovery has been extremely successful in fulfilling the human longing to understand the mechanical features of the universe, if for no other reason than to enhance survival. But science can answer only the "what is" questions, and comes up short on "ought," "purpose," and "meaning" questions. To understand this differentiation can go a long way in reducing tensions between science and religion. In the next chapter, additional focus will be given to the power of science in transforming human existence, offering up contextualized importance for the role the senses play in understanding the nature of reality, but also its limits in addressing purpose and meaning.

CHAPTER 4

Coming to Terms with Reality

Chapter Summary

Humans reside on a planet that is a biological oasis in many respects, given the surroundings of vast cosmological regions that do not seem all that friendly towards life. The more that is understood scientifically about how the pieces fit together and how the universe evolved to its present state, the more the word "amazing" seems to come to mind. Simply put, the universe has an astonishing amount of mathematical logic behind it. With this in mind, the discussion of this chapter considers varying models that address the nature of reality—1) arbitrary divine caprice, 2) divine sovereignty, 3) law based, or 4) indeterminate, with it being noted that there also exists a range of intermediate positions that may allow for a divine element on some level. Yet if reality is governed by law then it will be possible to discover that the natural order has emerged over time with a natural capacity to self-organize due to the laws of nature—and this, even though there may be random processes in play at the quantum level. This, of course, requires confrontation of long standing religious ideas about the theological nature of reality, where divine sovereignty is represented as God controlling all. The laws of nature would seem to suggest an alternative, at a minimum, a God who operates through laws and principles that allow indeterminate processes in nature to play out.

Scientific Context

It goes without saying that humans reside on an island planet that is a biological oasis in many respects, given the surroundings of vast cosmological regions that do not seem all that friendly towards life. The more that is understood scientifically about how the pieces fit together and how it evolved to its present state, the more the word "amazing" seems to come to mind. A while back I picked up a book entitled, *Is God a Mathematician?* The book does not directly answer the question represented in its title, but the author, Mario Livio, a physicist, sort of marinates in the apparent omnipresent and omnipotent powers of mathematics. He notes that in Galileo's view, and that of many others who have come after him, mathematics seems to be the language of the universe, offering breathtaking effectiveness at describing nature—from the motion of the planets to the makeup of matter, including the biological code of life. Simply put, the universe has an astonishing amount of mathematical logic behind it.

These insights have been rendered possible through the discovery that the natural order is governed by regularities that can be studied, analyzed and codified. What these regularities mean in practical terms is that methods now exist for projecting backward and forwards, processes now in operation. As to future events, projections can deliver reliable predictions (the most obvious being as noted earlier—the sun will rise above the eastern horizon tomorrow, even though readers will recognize the technical inaccuracy of that statement). Such understandings can also provide a basis for tracing processes back in time to learn a great deal about the past. These regularities are distilled to laws of nature, and ultimately mathematics.

Considering the sweep of human history, it is clear that knowledge is an emergent property as evidenced by an evolving understanding as to the function of the natural order. Such understandings often get reduced to mathematical formulation, marked by order and precision. By comparison, ancient times witnessed a far different prevailing view. Early mythology held that the forces of nature were the gods at work, and that the gods expressed pleasure and anger through these forces. That which was outside of human control or understanding was often characterized as supernatural and attributed to the action of the gods. This view has had a very long history dating to the Mesopotamian culture some 2000-3000 years ago, and earlier periods. But it was also found in early Greek culture, and interestingly some of these ideas still survive to this day in varying forms.[1]

---

1. For example, most readers will be familiar with the legal terminology, "act of God." This term has a history steeped in superstition.

Historically, important narratives circulated that reinforced the influence of deity in all aspects of human affairs. Some of this narrative got written down and preserved, with perhaps the most widely circulated being the Jewish Torah (the early books of the Old Testament). One of the distinguishing features of Jewish theology in relationship to the surrounding culture was that of monotheism (belief in one God), as opposed to polytheism (or many gods). But some of the ideas that came out of the Torah were in certain particulars not significantly different from that of surrounding cultural ideas. The Hebrew God was an active being that engaged in many spectacular actions. Perhaps the most spectacular example was the attribution of God putting the universe in place through fiat creation. This creation event was definitely not described as one having a natural explanation as read literally. Other attributions pitted the gods of the region against the God of Israel, with the end result of the narrative demonstrating that Israel's God was more powerful than the surrounding gods.[2] On those occasions when events unfolded nicely, the God of Israel was pleased, but when things did not go well, it was generally attributed to a lack of adequate faithfulness to God.

Many have covered at least some of this early history, and among them is Carlo Rovelli who was interested in identifying the point where the scientific method first met up with superstition. He identifies the pre-Socratic philosopher Anaximander from the period 610-546 BCE as the first historic figure that looked for logical explanations for events of the natural order rather than automatically attributing such events to the gods, or God.[3]

Yet even with this early beginning of the scientific method these old and very broadly held ideas regarding the intervention of deity in the natural order have not died easily. Many centuries after Anaximander the Greeks were still wedded to the idea of divine caprice as an explanation for the affairs of the natural order. Every aspect of life that was not fully understood was attributed to the gods. In fact, this idea extended to all aspects of life that seemed to be outside the reach of human control, including things like the weather, the success or failure of crops, and fertility of herds as well as survival-related events such as warfare and disease. It also extended to a great many other aspects of life—both the good and the bad.

Leonard Mlodinow discusses an interesting aspect of this a while back by tracing Greek development of mathematical theory. He centers attention

---

2. See 1 Kings 18; also, for a more in depth consideration of the Mesopotamian cultural influence on ancient Jewish text, readers may want to refer to the very accessible works of theologian Peter Enns, *Inspiration and Incarnation*, 2d Ed., Baker Academic, Grand Rapids, MI, 2015.

3. See generally, Rovelli, *The First Scientist: Anaximander and His Legacy*.

on the Greeks because they were instrumental in moving mathematics forward in a number of ways having to do with axioms, proofs, and theorems, but he notes that they had no theories dealing with probability. He suggests that one of the reasons for this may of had to do with their general belief that the future in all its detail (even wagers) unfolded according to the will of the gods, thus leaving no place for a theory of probability and randomness.[4]

Moving forward in history the idea of divine intervention in the affairs of nature continued to prevail, with heavy emphasis upon the miraculous and supernatural. Saint Augustine, for example, put forward the notion of "divine sovereignty," a model that viewed God as exercising complete and absolute sovereignty over the world and universe in all its minute detail. While this may sound appealing to mortals who seek security through the protective hand of God, it comes with consequences, not the least being the fact that if God is fully sovereign, then all that happens—the good, but also the bad—must be attributed to him, making it excruciatingly difficult to incorporate the Gospel message of a loving God. This approach, of course, overlooked the possibility that events might occur randomly based on the statistical laws of nature, including in some cases mathematical probabilities, and thus perhaps might have nothing to do with the whim of the gods, or God.[5]

Random Events and Divine Will

With this history as the backdrop for the prevailing worldview of the time, it should be clear that the early scientific community operated in a setting from which existed a strong headwind of mythology as a paradigm of how the world functions. Over time, as people discovered the constants and regularities that form natural law, progress was made in applying the scientific methods of observation and reason. These laws went a long way in adding order to human understanding of nature where it could be demonstrated logically that many occurrences attributed to the supernatural or miraculous have natural explanations. In the words of Kootsey regarding the scientific enterprise:

> " . . . The successes were frequent and spectacular enough to generate high optimism. These successes opened up a conflict

---

4. See generally Leonard Mlodinow, *The Drunkard's Walk: How Randomness Rules our Lives.*

5. Ibid., 27.

because there were now two sets of priests of the natural world: the old (priests of religion) and the new (scientists)."[6]

Even to this day, whenever events occur that are unusual, it has been common for some to attribute it to the supernatural. Reader may recall a few years ago that televangelist Pat Robertson was blaming a nasty hurricane on gays, and atheists. The obvious candidates for the category of supernatural intervention generally involve the forces of nature that either accrue positive or negative benefits—including things like being struck dead by lightning, death or survival of a killer earthquake or tsunami, floods, and other occurrences of such genre. One of the great contributions of science and mathematics is that it has provided a way of differentiating natural from supernatural. As Kootsey points out, a street definition of supernatural would be "beyond or outside of the natural."

There has been a long tradition of associating natural events with the supernatural, yet by studying the history of past events that have been tagged as "supernatural" it is possible to discover that most events can be explained today as being within the realm of the natural world, with natural explanations. Certainly most recognize that a great many of the so-called "acts of God" have very natural explanations that the ancients were not prepared to understand. Yet, the problem is elusive as Kootsey points out:

> What is natural is derived from human experience, so that supernatural also means outside of human experience. Applications of this intuitive classification have repeatedly been superseded. For millennia, it seemed clear that humans could not fly. Such behavior was only possible for birds, insects, and (supernatural) angels. Now that limitation seems quaint. Until recently, communication between humans was only possible when they were in direct sight or hearing of each other. Now we can communicate instantly with anyone on the planet, utilizing technologies firmly based on well-known and understood natural laws . . . Even if we had perfect knowledge of natural laws, we can and do have very incomplete knowledge of what can be accomplished through those laws. The scientific identification of the truly supernatural thus depends on our current level of knowledge and technology. As a consequence, supernatural events do not constitute a reliable method for establishing belief in God.[7]

---

6. See note 22.

7. Kootsey, http://atoday.org/ultra-darwinists-and-conservative-christians-agree.html.

To this day it is not uncommon for unexpected events of the natural order to get labeled as part of the divine will or action, presupposing that God intervenes routinely in the affairs of nature. Yet, if this were in fact the case, the scientific method would not work because it would produce inconsistent results due to the intervention. But to the contrary, not only can an ever-increasing amount of what happens in this world be explained scientifically, but also quite often it can be done on mathematical terms with mathematical precision. Even in cases that cannot be reduced to some direct deterministic equation, reality is still governed by probabilities, such as where some random occurrence falls within a probability distribution, producing an unusual but predictable outcome, calculated with a mathematical degree of confidence. In some cases, when the worst possible outcome happens, it is possible to conceptualize other possible outcomes if the participants to the event had altered their schedules in some seemingly insignificant way. The point is that being a factor in a probability analysis does not destine a person for a given fate, but random events can sometimes align to make these outcomes possible. This is what makes life so unpredictable.

It may be useful to consider some examples at this juncture. It is possible, for example, to discover the epidemiological probabilities of acquiring a specified disease given certain relevant variables—genetic predisposition, lifestyle, geography, etc.; or the sociological probability of becoming a crime statistic—age, race, location, time, etc. In all such scenarios, if the samples are drawn properly, and there is an adequate accounting for the variables, such determinations can be made with a mathematical level of confidence. Thus, where seemingly small insignificant factors and random events align to make an outcome possible, the underlying statistics can generally be anticipated to be within the range of a normal mathematical distribution.

Interestingly, human understanding of statistics has advanced to the point that it can be used to uncover supposed random patterns that defy logic, requiring an explanation outside of random processes themselves. In games of chance, for example, such a pattern can quickly elevate cheating as a suspicion for anomalies that fall outside the range of a normal distribution. Analysis of statistical probabilities and standards are in fact used forensically to uncover criminality.

So in considering the range of events that can accrue in the normal course of things it is terribly interesting that mathematics plays a key role. It has explanatory power for the regularities that exist in nature that can be studied, analyzed and codified—even in cases that are deemed indeterminate. In practical terms, these regularities afford a method by which processes in play can be projected to make reliable predictions about future events. Likewise, processes can now be traced back in time to learn a great deal about the past. These regularities are distilled to laws of nature, and ultimately mathematics.

All of this would seem to reinforce a deterministic understanding of reality. In the popular mind there may be a tendency to merge the ideas of divine caprice and divine sovereignty, yet there are distinguishing characteristics—not so much in results, as it would be in process. Caprice denotes arbitrariness, which would surely be something that could be identified mathematically, while divine sovereignty conveys the sense of a slightly more sophisticated understanding of divinity that incorporates the idea of rationality and order, with its main emphasis being that God controls all. Those who are more inclined to recognize the power of science and mathematics, have yet a third approach—that being a form of determinism that makes no appeal to a divine.

Ultimately, there are sound reasons to believe that God-fearing individuals are not excluded from the laws of nature in some way different from their secular counterparts. Bad things do indeed happen to good people, and the story that mathematics would seem to tell is that there is little room for reading any theological meaning into tragedy—in terms of divine will. When an individual is accustomed to praying for and expecting the protective hand of a God who controls all, there may be psychological solace in processing the untoward events of life, but the proposal being suggested here is that these are the product of statistical probabilities and associated random events that do not necessarily have anything to do with God's will. To the extent this is the correct conclusion, conceivably the human prayer should not be the seeking of divine intervention from the laws of nature, but strength to deal with life's vicissitudes, and gratitude for those things that make life supremely wonderful.

Indeterminacy

In addition to the varieties of determinism, there also exists another category—something briefly mentioned above—that being indeterminacy. It is found under a variety of circumstances, including situational outcomes that are dependent upon a number of variables that can interfere with a planned or desired outcome. But first and foremost this principle is thought of at quantum scale where subatomic particles behave in ways that cannot be determined ahead of time. As its description implies, an individual outcome of a process are indeterminate, even though they do represent a modified form of determinism in that they offer results within a probability range, much as a large number of trials for a coin toss will result in approximately half of the tosses being heads and half being tails, even if an individual trial is rendered an event with an indeterminate outcome.

It is the creative thought of the late Ian Barbour, who was both a theologian and physicist, that indeterminacy in nature provides an avenue for thinking about the possibility of a moral universe and a world populated by moral beings, as opposed to deterministic robots.[8] Barbour's suggestion is intriguing, though reality would still seem to come with a modified form of determinism based on the laws of probability.

Summary Thoughts

Perhaps at this juncture it may be of value to summarize some of the ground covered up to this point. Several Chapters back discussion was given to a general framework for how knowledge is acquired, and one of the primary ways in which this occurs is through the methods of science. For that reason, it was important to do an actual review of the methodology of science. With its rise as part of the discovery of natural laws and principles, humans have learned that many aspects of the natural order are explainable and predictable, and not the product of divine caprice.

Also in a prior chapter brief mention was made of the fact that the senses can inform by providing knowledge of *what is*, though it may not be very good at addressing *ought* type questions, or to provide a context for meaning. *Ought* questions, then, becomes the area of knowledge that derives from revelation, and will be considered in future chapters—it coming with its own challenges.

In this chapter four understandings of reality have been outlined—1) divine caprice, 2), divine sovereignty, 3) law/mathematical based, and 4) indeterminacy. This latter category is noteworthy in as much as it may allow for a moral order on some level, recognizing the natural order to be the product of a universal process that has emerged over time with a certain natural capacity to self-organize in spite of random processes in play—most significantly at the quantum level.[9]

Certainly for those who understand God as "creator," to consider the thought that random process exists within that creation has enormous

---

8. Ian Barbour, who was both a physicist and theologian, suggests in his book entitled *Religion and Science*, the possibility of divine interaction at the level of quantum physics, given the law of indeterminacy. This, is, itself a fairly complex discussion somewhat beyond the scope of this book. It is mentioned here, though because some readers may wish to investigate this idea further.

9. Ibid.

theological implications. This is particularly the case for those nurtured on Augustinian thought, for the existence of randomness implies that *probability* plays a role in the processes of the natural order and in human experience.

One of the challenges for Christians and those of other faith traditions is that the transcendent reality to which faith speaks can seem very real to the mind, and yet there are so many versions that it quickly becomes obvious that certainty must loom as very elusive. This becomes particularly evident when contrasted with knowledge acquired through the senses. Thus, for those who argue that knowing comes through revelation, the utter breadth of the many claims to revelation, as well as their interpretations, raises the question of the criteria by which such claims will be assessed. Of course, some give no attention to analytics, and to the extent that this is true the query emerges as to what credible value can be assumed for certitudes of mind devoid of some foundational considerations.

This issue will be considered to some extent in the next chapter, and it must be evaluated by looking at some of the problems that are encountered in the process. Certainly anyone who opens the pages of Scripture for the first time will be overwhelmed with its complexity—its many authors, obscure meanings, conflicting themes, distance in time and culture, etc. However, in most Western countries for those reared on the importance of Scripture it is often understood as God's revelation, and anyone who comes from a Biblicist heritage likely has been handed a set of beliefs derived and distilled from it, and thus what might otherwise be overwhelming has been made simple by a set of doctrinal statements that attempt to extract the essentials. Yet, problems remain, and some of this will be taken up in the next chapters.

CHAPTER 5

Human Frailty: The Hidden Face of Idolatry

We like to build things . . . We construct our house of notions and expect to inhabit them for extensive periods of time. We step inside our newly established structure, close the door and isolate ourselves.[1]

Chapter Summary

Because of the existential insecurities all humans endure, there has long been an effort to wrap reality in garb that is impermeable to earthly truths. As such it is not uncommon for the claim to be made that all knowledge comes from revelation, and if this is the correct view then outside knowledge cannot flow into Scripture. It will always be seen as the other way around, often supported by the mantra of sola scriptura. This chapter gives consideration to some of the more obvious human frailties that surface with respect to understanding reality—frailties that consistently tend to elevate the authority of certain identifiable elements of a belief system well beyond an appropriate epistemological context. This includes idolatrous veneration of sacred writings, doctrinal authority, institutions, personalities, and traditions. A superficial reading of this chapter may leave the reader with the impression of intent to discredit these symbols of sacred authority, though the actual objective has been to question some of the prevalent assumptions (inerrancy, infallibility, possession of "truth," etc.) that are not well supported in the real world. Yet it is possible to harbor a small amount of compassion

---

1. This is an excerpted adaptation by Gabriel M. Riojas, "A House of Idols," *Spectrum Magazine Blog*, http://spectrummagazine.org/article/gabriel-m-riojas/2013/09/30/house-idols-introduction, 2013.

towards those who elevate these symbols well beyond an appropriate level, for if the human condition is taken into account it should not really be too surprising that misplaced authority should become a refuge for some people in a hostile and uncertain world. Such behavior should probably be seen as a human coping mechanism that has power to sooth anxieties—perhaps acting as a symbolic fig leaf to cover existential nakedness. There is power in reading a set of words and characterizing them to be "God's Word," or an even more powerful characterization, "God's inerrant Word," or that certain humans or institutions have discerning infallibility. To understand that this is why many gravitate to narrow and extreme approaches to faith allows for more charitable expressions towards those who proceed in this manner. Nevertheless, there is strategic value in recognizing what is going on here, and avoiding its seductive influence.

In the preceding number of chapters, consideration has been given to the question of how humans come to knowledge, with a great deal of focus on the strategic importance of the senses, moderated by a systematic use of reason. But there are other modes of knowing and it is now time to expand the focus—that is, to consider the role of authority. There are various forms of authority, and among these is revealed authority. Those who hold such beliefs see knowledge as coming through divine revelation conveyed through sacred texts, with it representing an important ingredient towards development of a worldview that transmits meaning.

Yet, because of the existential insecurities humans endure, there has long been a tendency to make assumptions about reality that cannot be validated, and in some cases are even falsifiable. For example, it is not uncommon for the claim to be made that all knowledge comes from revelation, and if this is the correct view, then outside knowledge cannot flow into Scripture—it will always be seen as the other way around, perhaps supported by the mantra of *sola scriptura*, or by Scriptures alone. Such views understand the Scriptures to be its own interpreter, not subject to outside sources, such as those relied upon by research methods.[2] If it were otherwise there would be openness to the findings of anthropology, sociology, history, science, as well as literary analysis.

The problem with such insularity is that there then is no truly objective basis by which to study the many and varied so-called revelations that

---

2. This is the area of biblical study referred to as hermeneutics, discussed briefly later in this chapter.

have surfaced down through history—including revelations that are sourced outside of the Judeo-Christian Scriptures. Shorn of the capacity to conduct an objective evaluation, there is no basis to ascertain the validity of one revelation over that of another. If it were otherwise, life would be a whole lot simpler.

With the foregoing in mind, this chapter will give consideration to some of the more obvious human frailties that surface with respect to this category of knowledge—frailties that consistently tend to elevate the authority of certain identifiable elements of a belief system well beyond an appropriate epistemological context. The first of these to be considered has to do with the dominant sacred text used by Christians—namely the Bible composed of the Old and New Testaments.

To illustrate this point, a while back I posted comments on a public forum regarding a proposed change of wording in the Statement of the Fundamental Beliefs on the Genesis creation narrative by the denomination of my birth. The proposed wording change would tend to move the statement from a status quo that is scientifically neutral, to one that is impliedly anti-scientific.[3] My brief comment stated that it was my hope that the committee considering the wording change would include prominent representation from the science community. In response, A. Mitchell, whom I don't know, posted a reply comment as follows:

> Our mandate is from God through His Scripture. We are to represent His way as He directs from His book . . . If God indicates something in His word, whether it conflicts or agrees with science; whether or not it seems reasonable, we accept it. God is the arbiter of ALL truth. His word is absolute. Scientific explanations and social mores must bend to scripture, not the other way around.

These remarks will likely resonate with some people, but it will be my argument in this chapter that such a view represents a metaphorical structure that has all the windows and doors shut tight, and as such expresses a profound misappropriated use of Scriptures, particularly in view of other avenues of knowing.

Within Christianity it is important to be clear about what is often taking place—frequently Scripture is elevated and promoted as "God's word," and held by many to be inerrant—either in belief or practice. For those who hold this high view, it is difficult to engage in objective conversation, for the starting presupposition prevents it. In the face of statements like the one above, objective considerations become impossible.

3. Including verbiage that creation was recent, on the order of a few thousand years ago, and that creation occurred over six 24-hour days.

With this having been said, I must admit that I do harbor a small amount of compassion towards those who proceed in this fashion for if the human condition is taken into account it should not really be too surprising that misplaced authority sometimes becomes a psychological refuge for people living in a hostile and uncertain world. Such behavior should probably be seen as a human coping mechanism that has power to sooth anxieties—perhaps acting as a symbolic fig leaf to cover existential nakedness. There is power in reading a set of words and characterize them to be "God's Word," or an even more powerful characterization, "God's inerrant Word." So if it is understood that this is why many gravitate to a fundamentalist approach to faith, it allows for more charitable expressions towards those who proceed in this manner.

In considering an appropriate way of understanding Scripture, it may be helpful to refer back to the discussion in chapter 1 on *certitude* which looked at the ways by which finite humans acquire knowledge. It was there that a framework was developed for prioritizing the certainty that is brought to the thinking process. It may be recalled from that discussion, that there exists a vulnerability with all faith systems—Christian or otherwise—which attempt to assert an unassailable authority for it sacred text. The problem, of course, arises in establishing objective criteria for evaluating its merit in the first place. In fact, the argument I will be making is that some of these approaches give the appearance of idolatrous constructs.

Foundational to any discussion that touches on the question of "idolatry" would be an examination of how the term is being used—particularly since it has historic roots in monotheism as a referent towards other mythologies—with it applying to the gods of some outside tribe, rather than the one true God. But the larger context of meaning for this term includes unwarranted focus on things like money, power, and pleasure. As used in this chapter the term addresses exaggerated claims with untoward acts of reliance on a variety of authorities—including Scripture, doctrines, institutions, personalities, and traditions. Once it is recognized that there is a potential for misappropriation of these categories, such use can often fit the definition of a lower case (g) god. Some specific examples will be briefly examined below, though not in any exhaustive detail. The intent is to put enough information into this chapter so that anyone wishing additional resources for further research will be able to do so.

God's Inerrant Word

Humans do not spring from their mother's wombs believing the Bible to be God's word, much less that it is inerrant. This path is often by a foundational process that includes a measure of family and cultural tradition. So the question becomes one of how humans ascertain the existence of reality beyond family and culture?

I would like to propose an approach to all sacred writings that is respectful, yet at the same time do so in a way that keeps it real. To start with, none of this discussion is intended to delegitimize the Judeo-Christian Scriptures, but rather to question the point at which an inappropriate level of authority is vested in it either by way of assertion or practice. Perhaps one way to consider this issue is within the context of the larger world beyond the Judeo-Christian tradition. It is noteworthy that a large percent of humans in the world do not share this tradition and not surprisingly have come to different conclusions regarding biblical authority. In pondering this reality, it is important to understand that *tradition* has played a key role, and even though it is possible for people from one tradition to convert to another tradition—in general this is not the common path to belief in Scripture as foundational to faith and practice.

Some may be inclined to blindly acquiesce to childhood inculcation or social traditions—never to investigate the details in any further depth. But as important as tradition may be, most thoughtful people would likely argue that the authority of Scripture should extend beyond tradition alone since its source may not be self-evident. Furthermore, if this matter is to be resolved in any sort of satisfactory fashion, it must be connected to the real world of evidence. That means, for example, that if sacred writings are to be considered as "God's inerrant word," either in belief or practice, it will be necessary to resolve the problems that such a construct creates.

Occasionally in order to get at a reality it is useful to turn to analogies, and earlier some consideration was given to the difficulty in approaching the sacred on objective terms, and with this in mind it may be worthwhile to momentarily consider a completely different religion from Christianity—namely, Islam. Islam seems a good contrast to Christianity for the simple reason that the way in which it approaches its sacred text, the Qur'an, can be instructive as to the way some Christians approach their own sacred texts. The dominant view within Islam regarding the Qur'an, which for Muslims is considered to be the word of God, is that it is infallible and inerrant, understood as perfect in every detail as read in Arabic, as well as internally validated. The significance of this last point is that for anyone wanting to investigate the Qur'an as a divine source of truth, it must be read and studied, and its authenticity will be revealed within a reading of the text itself. As may be recognized, this approach is not dissimilar from what Christian fundamentalism engages in when it asserts the Bible is self-authenticating.

What should be obvious is that a huge barrier exists between the Muslim and the non-Muslim. To enter any sort of meaningful dialogue with a believer of the Muslim faith by a nonbeliever, where its primary faith document is internally validated, it will become an obstruction that must be overcome if effective dialogue is to be accomplished. This obstruction is compounded for any Christian who likewise proceeds on the basis of an internally validated scripture, on the order of s*ola scriptura*. Any possibility of dialogue between the Muslim and the Christian thus requires a quest for rules of engagement—rules that can bridge such barriers—that would be foundational to the possibility of any meaningful dialogue. The tools of this sort will be familiar to most readers, for they are those of the historical-critical method—analysis that looks to both internal and external data based on science, history, anthropology, and a number of literary factors.

Previous mention was made of the bumper sticker with which most readers will be familiar, "God said it, I believe it, and that settles it." It is interesting to observe that fundamentalist of any monotheistic religion—Judaism, Christianity, or Muslim—could carry this bumper sticker as representing an approach to their respective sacred Scriptures, its message suggesting that the adherent passively soaks up a reading of the chosen sacred text like a sponge, truth being uncomplicated and rather obvious with no authentication of the text being necessary. Yet to proceed in this manner suggests a rather naïve process that ignores many important considerations, including the—who, when, what, and where.

The degree to which "believers" are truly open to the challenges of a sacred text such as the Bible is largely a function of what is allowed in as evidence—that is the hermeneutical approach. This is, in fact, the point of departure that has long separated fundamentalists from the more mainstream and secular approaches having to do with the threshold question of whether Scripture is to be judged exclusively by its internal merits (the Bible alone), or whether it should be based on a more neutral and objective standard? With this in mind, it may be useful to give more scrutiny to this matter, as there are two very different models that are frequently used. They are roughly as follows:

The Importance of Hermeneutics

Hermeneutics has to do with the rules by which the Scriptures are to be read and interpreted. In this regard I would like to briefly outline two prominent approaches. First in this regard is an historical grammatical hermeneutic which attempts to discover the meaning of a passage by seeking the author's intent, as well as to discover how contemporaries would have understood it. It diverges from a more thoroughgoing process that analyzes beginning assumptions. The critique of this approach relates to its refusal to include the full range of objective criteria for analysis, part of the common ground that is fundamental to the acquisition of knowledge. The question arises as to how a religion—Christian or otherwise—can claim a commitment to truth if it removes from the equation any critical review of its primary documents. The second of these approaches is often referred to as an historical critical hermeneutic which does not presuppose Scripture to be God's inerrant word, but that there is a unity of all reality, and that objective and neutral criteria should be the basis of inquiry. When considering a very different faith-community from Christianity—like Islam—most Christians will see the appropriateness of operating with neutral objective criteria, and that such would be the common ground by which dialogue could take place with a Muslim. Quite simply, it is the most objective standard by which any claim to revelation can be evaluated by a skeptical outsider. The critique by many fundamentalists revolves around a willingness to begin with more open-ended starting assumptions.[4]

The methodology used is significant because it will influence the handling of conflicting narratives, interpretations of the text, and manuscript variations. Furthermore, how does one determine that it is "God's Inerrant Word" or more simply "God's Word," or should the reader just make that as a starting assumption without any attempt to validate the claim? This is a foundational issue that is more difficult to answer than what some may be willing to acknowledge. Of course, if this is the starting assumption, as may be the inclination of the historical-grammatical-approach, it may be seen as a non-issue. But on the question of inerrancy, it is not so easy to overlook this assumption, for it can be internally tested.

In light of the discussion in chapter 1 on "knowledge" and the process by which the certainties of the human enterprise are advanced, the answers to some of these issues should not be taken for granted. The perspective regarding inspiration cited above captures an important element of how many people read and understand Scripture—it being simply a tendency to ignore

---

4. See, for example, Richard M. Davidson, "The Authority of Scripture: A Personal Pilgrimage," *Journal of Adventist Theological Society* 1/1 (1990) 39—56).

basic questions and to believe and trust that they have captured the correct understanding without critical analysis.

The thesis of this chapter is that it is appropriate to think of the Bible in sacred terms in as much as it attempts to connect the human and divine. But it is important to also recognize that humans authored the Bible, with them attempting to say things they found meaningful, yet which in the end heavily contains the imprint of humanity. As such, the Bible should not be viewed as the "word of God," and much less the idea that it is inerrant. This is not to suggest that God cannot be found within its pages, nor should it exclude the possibility of him having influenced some of its authors. What seems clear, however, is that many have elevated Scriptures to such an unwarranted level of authority that it has become an object of worship by way of putting forth unsustainable claims.

In most cases, it would probably be charitable to hold out that many arguments are made that have a purity of intentions, even if misguided. This chapter will take a look at some specific reasons why such constructions are inappropriate. That said, there is little doubt but what the set of books that compose the Judeo-Christian Scriptures represents a rich treasure of documents that detail the history of the human longing for a purposeful existence, seeking a connection with the divine. This motive seems inherently woven into the fabric of its many books.

So with this little preamble, perhaps the way to begin is to dwell on some central themes. First, let's consider the theme that emanates from the New Testament. As we consider this, most will likely agree that in general it portrays a God of love, this being a central premise. Yet, this is but one theme found in Scripture. Another theme from the Torah (first five books of the Old Testament), does not necessarily lead to the conclusion of a God of love, particularly when isolated from the dominant perspectives represented in the latter portions of Scripture. Sometimes represented in the Torah are expressions of an angry, vindictive and arbitrary God. Because of these inconsistent portrayals, readers are left to ponder why God's "alleged" inerrant word would shift in such a significant way, unless some of the ideas represented were not inerrant.

A question of some import is whether there is truly a fundamental difference in perspectives going on or whether Scripture is part of a unitary theme? If the latter viewpoint is to prevail, then apparent discrepancies are not what they seem. A unitary theme would be an argument of necessity for anyone claiming Scripture represents "God's inerrant word," for if it is inerrant, then every portrayal of God must be deemed equally valid, thus the need for explaining away apparent conflicts.

Sample Problems

In order to get some sense as to the magnitude of the problem there would likely be some value in taking a look at some of the more difficult passages, and in doing so it would be well to keep in mind what is arguably the central theme of the New Testament—a God of love.

God kills and Commands his follows to Kill and Commit Genocide[5]

- When the king was stubborn and refused to let the Israelites leave the land of Egypt, the Lord killed every first-born male, both human and animal. Exodus 13:15; Anyone who works on the Sabbath day, or gathers firewood on the Sabbath is to be put to death. Exod 31:15; Num 15:32–36
- Aaron's sons did not present God with an offering of holy fire, so God struck them dead. Lev 10:1–2
- God commands that any child who is stubborn, rebellious, or curses his parents is to be stoned to death. Lev 20:9; Deut 21:18–21
- God commands that anyone who commits adultery or has sexual intercourse with an engaged person is to be stoned to death. Lev 20:10: Deut 22:23–24
- God commands that anyone who consults with the spirits of the dead is to be stoned to death. Lev 20:27
- God will deliver into the hands of the Israelites the lands occupied by others. The Israelites are to kill the entire population of their enemy; they are not to make any alliance with them or show them any mercy. Deut 7:1–2
- Anyone who curses is to be stoned to death. Lev 24:13–16
- God orders the slaughter of men, women & children for the sins of their ancestors. 1 Sam 15:2–3

God Discriminates Against Those Who Have Physical Defects

- God commands that no one with a physical defect—blind, lame, disfigured, crippled, etc.—can make an offering to the Lord. Lev 21:16–21

---

5. Havener. *God on Trial: The Verdict*, https://www.youtube.com/watch?v=dx7irFN2gdI, a dramatization of some of these problem that can arise from a non-discriminating reading of Scripture.

HUMAN FRAILTY: THE HIDDEN FACE OF IDOLATRY

God Teaches the Principle of Retribution

- God says the principle shall be a life for a life, an eye for an eye, a tooth for a tooth. Lev 24:18–22

God Approves of Slavery/Sex Slaves

- God sanctions slavery Deut 15:12–18; 20:10–11
- God commands Israel to kill the Midianites; to kill all the men and women—anyone who has had sexual intercourse, but the young virgins can be saved for use by the Israelites. Num 31:1–18

Bad Things Happen to Those That Disobey God

- The Lord commands the Israelites to worship God exclusively, or they will be destroyed completely. Deut 4: 23–26
- If you reject the Lord, or do not obey his teachings he will strike you with incurable disease, send drought and wind to destroy your crops. Deut 28:15–24; 58 61

God Rejects Those Who Are Victims of Circumstances outside Their Control

- No one born out of wedlock or any descendant of such a person, even in the tenth generation, may be included among the Lord's people. Deut 23:2

God gives Polygamy Implied Approval

- God instructs that individuals should not disgrace their fathers by having sexual relations with any of his other wives, but meanwhile his instructions make no mention of the practice of polygamy itself. Lev 18:8

The Message of Jesus in Contrast

- The greatest commandment is the love God, and the next is akin to it—love your fellow man. Matt 22:36–39
- Do unto others, as you would have them do unto you. Matt 7:9–12; Luke 6:27–31

- You have heard it said "an eye for an eye, a tooth for a tooth, but I say unto you, take no revenge. Matt 5:38–40
- Do not judge others. Matt 7:1
- God is Love. 1 John 4:16

Well, what should be made of these contrasting passages, where the New Testament views are put up along-side pre-New Testament portrayals of God sanctioning crusades of slaughter, and certain unethical practices such as slavery? Could this have been a moral God of love? More importantly, how would these divergent perspectives be reconciled, particularly if these passages represent God's inerrant word?

As suggested above, many who support inerrancy of Scripture attempt to read the Old and New Testaments seamlessly, where all of Scripture is viewed to be in complete harmony in every detail, with no particular priority given to one passage over another. Both Testaments are understood to be the "word of God," and as such, are not in conflict. Under this view, the many representations of God in the Old Testament are merely the other side of the same loving God of the New Testament. Those taking this approach understand this loving God to have the right to pass judgment on people along the way, with the divine perspective deemed to be good and just—in fact, God, as God, can do as he pleases.

Sort of related to this discussion it should be mentioned that Bart Ehrman, a Princeton educated theologian, wrote a book recently titled, *God's Problem*, doing so on the important question of why humans suffer. He notes that the dominating theme that comes out of the Old Testament is of a reality where God rewards those who worship and obey him, and punishes those who fail to live up to his standards. In view of this, should it therefore be understood that those who suffer misfortune are receiving divine retribution? In certain important ways, this thesis is certainly evident in many of the passages referenced above. This issue is certainly worth reflecting on, and more will be said about it in the next chapter.

As a kid I was always somewhat troubled by some of these ideas. For sure, the doctrine of inerrancy sort of requires the reader to adopt the conclusions of all Biblical writers without any discriminating filters. So to the extent that the reader is locked into a narrow understanding of the nature of revelation there may be little interest in considering other possibilities. The point above, however, is to highlight and illustrate the difficulty inherent in human understandings of revelation.[6]

---

6. While I was not reared within a sectarian tradition that formally subscribed to the doctrine of inerrancy, it is quite clear that many who deny inerrancy formally,

Over against an approach of inerrancy, there are a couple of other ways in which to systematically address the dilemma of differing perspectives of the divine character that surface. The first of these was put forward by the second century Christian author, Marcion, who argued that the Hebrew God was different from the Christian God. He noted that many of Jesus' teachings were incompatible with the actions of the Old Testament God, and for these reasons lobbied for Christianity to recognize the Jewish God to be a different being than the one that ruled Christianity. However, as most reader will know the flow of Christian history did not adopt this understanding. This, then, leaves the possibility of a third approach, which takes the earlier documents, that is, the Old Testament, seriously by viewing it through the prism of a more human dimension. In short, it recognizes the biblical authors to have written from a very human perspective, with humans by nature comprehending reality but dimly. If it is recognized that understandings emerge over time it should not be terribly surprising that moderns have a different concept of God than did Mediterranean nomad of 2500–3,000 years ago. Those inclined to reject this latter approach must explain why Jesus said so many things contrary to some of these ancient traditional understandings.[7]

More Trouble for Inerrancy

In this section consideration will be given for other reasons to navigate away from the doctrine of inerrancy. For this, it may be instructive to turn once again to New Testament scholar, Bart Ehrman, who tells the story of his journey from fundamentalist Christianity to agnosticism. There are hints that some of his fundamentalist presuppositions actually contributed to this outcome. He attended Dwight Moody Institute, where he studied biblical languages, and then went on to Wheaten College to complete his undergraduate study. Thereafter, he transitioned to Princeton Theological Seminary to complete both a Master of Divinity and Ph.D. programs. He conveys that when he arrived at Princeton, he was very much committed to inerrancy on the question of inspiration, but he eventually discovered that the data simple would not support this presupposition and this eventually led to a crisis of faith.

Ehrman's knowledge is sufficiently well developed regarding the evolution of the Canon that it would seem to be worth taking a look at some of

---

embrace the presuppositions of inerrancy in practice.

7. What comes to mind here is the Sermon on the Mount where Jesus tells those assemble, "you have heard it said of old . . . but I say unto you . . ."

the evidence he discusses—evidence that is common knowledge within academia, but not well known at the pew level. The points that will be considered do not so much attack the credibility of Scripture itself as it does to cast a rather sobering reality check on claims that are sometimes put forward.

Perhaps one of the preliminary considerations that should be made is to recognize that not any of the original manuscripts exist. This becomes significant because each of the earliest manuscripts have variations, with none being completely identical. While many of these variations appear to be the result of inadvertence in the copy process, and in many cases rather insignificant, this is not always the case, with some impinging upon doctrinal considerations. The primary point to be made here is that it is difficult to determine which among them is the closest to the original text. If, in fact, the originals were inerrant, with this being a central construct of how God expects revelation to be understood, it seems rather strange that he would allow the original documents to all disappear, leaving instead copies that contain variations and material discrepancies. If inerrancy were a position worth defending, would God not have made certain that the original versions were preserved so as to negate error—if they were in fact inerrant in the first place?

Bart Ehrman and others have documented a number of points, including the following:[8]

The copyists—the only way to reproduce a manuscript in the ancient world was to do it by hand, letter by letter, one word at a time. While today the expectation is that every copy of a book will be exactly the same, this was not so in the ancient world. All manuscripts, including the Bible, were each one of a kind with no two copies being exactly the same. Anyone reading a book in antiquity could never be completely sure that they were reading exactly what the author had written. Copying texts allowed for the possibility of manual error; and the problem was widely recognized throughout antiquity. The writing technique included the practice of all lower case letters with no punctuation and no spacing between words. Ehrman notes, for example, a text might appear like the following: [godisnowhere], and obviously this could mean quite different things, as in, a) God is no where, and b) God is now here. This form of writing, called *scriptuo continua* is illustrative of the issues confronted by all copyists in interpreting the intent of the original author.

---

8. See generally B. Ehrman, *Misquoting Jesus*; for a slightly more scholarly treatment of this subject, see generally T. Stark, *The Human Faces of God*.

Many of the earliest scribes were not professionals—therefore mistakes were more commonly made in transcription, and more frequently because of inexperience, and general lack of professional training.

Mistake/negligence—this is the biggest category, where scribes made copies from manuscripts that already contained mistakes. So they ended up copying the original mistakes and then added to new manuscript additional errors of commission or omission so that over time the errors tended to accumulate, compounding the inaccuracy of later editions compared to the original text. In this way the mistakes multiplied over time. Here is a list of some of the errors to be found as identified by Ehrman, errors about which he goes into some documented detail:

- Spelling errors
- Duplicated line errors
- Sound alike word errors
- Mistakes of nonsense vs. sense— "sense" mistakes are more difficult to handle in as much as they appear to have been done with some volition, while the nonsensical variety can be counted as simply human error.
- Intentional 'perverse audacity' altering text to make it say what the scribe wanted it to say.
- Additions/deletions/misspellings, other blunders
- Some alterations made with intent at clarity
- Corrections for what was believed by the scribe to be factual error
- Interpretation errors
- Theological motivations for alterations
- Over time new bogus gospels were introduced by individuals who would write in the prominent name of another, i.e., *The Gospel of Thomas*, attempting to fool the reader into believing the author to be someone it wasn't.
- Total error estimate, old estimate—eighteenth century—research by John Mill, Queens College, and Oxford—found 30,000 variations among 57 surviving manuscripts in the Greek; New estimate—at last count over 5,700 Greek manuscripts that have been discovered and catalogued with 200,000—400,000 variants identified[9]

---

9. Ehrman, 84–85; 88.

Theological significance of manuscript variations—Ehrman goes into some detail about the controversy over the most explicitly Trinitarian passage, 1 John 5:7—8 and the Johannine Comma, which represents a more explicit description of the Trinitarian doctrine that was inserted into the Latin Vulgate, but which is generally missing from a majority of the Greek manuscripts. This tends to weaken the authority for the doctrine, since it is built to a large extent on this added passage.

This chapter was initiated with the purpose of identifying areas where religious people put certain ideas at the front of the line of importance that would seem to supersede all else; ideas that may support a narrative that they may have a vested interest in, but which rests on rather shaky intellectual grounds. The term that has been borrowed from the religious community is *idolatry*.

One of the major ideas in this genre of discussion has been the doctrine of inerrancy. In the discussion above, a number of specific reasons have been presented as to why this doctrine deserves to be rejected, both intellectually and in practice. Certainly each of the points discussed above demonstrate the weak foundation upon which this understanding of scripture was built—in spite of its prevalence. In fact, more than anything its prevalence would seem to underscore the deep human longing for a strong and reliable anchor, and for sure it seems that this doctrine succeeds on these grounds. Yet it is not backed up well by evidence. This means that formally embracing this doctrine, or its more dishonest version, which is to deny this doctrine yet to go on and treat Scriptures as inerrant, is inappropriate and comes at a cost, for when discovery is made that the reality does not match the idea, it can create a spiritual crisis for those who have a vested commitment in maintaining a mythological system that does not stand up well to the evidence.

This discussion has focused in certain ways on substantive points that can sometimes impact doctrinal understandings. It can also impact how Scripture is read, whether as equal weight to all parts of Scripture, or whether it is read strategically, with it having a certain trajectory. However, for the most part, this discussion has been procedural in nature, attempting to shed light on some often overlooked but important procedural issues that can influence how inspiration is to be understood. If nothing else this discussion should convey the fluid nature of human understandings and a certain folly in holding to rigid and inflexible positions. The human longing for certitude in life may lead to a benign neglect of some of these important details, or they may be explained away as inconvenient to settled and cherished views that might otherwise dislodge certainty.

Doctrinal Authority

In addition to the tendency of some to elevate the authority of a sacred text such as Scripture beyond what the evidence would suggest is appropriate, there are other unhealthy ways in which humans engage in this conduct. One possibility would include unhealthy attachment to a set of beliefs sometimes referred to as "doctrines," where a given construct becomes a definitive and intractable part of a narrative, with no amount of new data becoming a point around which consideration will be given of the possibility that a doctrinal revision is in order, or that a modified paradigm should be contemplated. There is a human tendency to commit to one's own prejudices, and what this means is often there is an inclination to not be open to other possibilities.

With this in mind, I found the following response to an article I published awhile back to be helpful. One reader submitted the following: "Do you suppose it's possible to convert 'I shall not be moved, even if I'm wrong' to 'Please help me to move if I'm wrong'?" This certainly seems to epitomize the healthy spirit in which the human quest should operate.

Institutional Ghettos

Around many ecclesiastical institutions grows cult like followings. It appears to often be a two-way street between institution and parishioners, with the institution many times claiming to be the one and only gateway to God; the one true church or other similar framing where such claims can suck in those inadequately equipped to think for themselves. For the institutions and the powerbrokers within it, the obvious incentive for fostering unhealthy attachments can be—the financial benefits that can accrue through such attachments—membership growth, followed by growth in charitable contributions, trusts and other gifts. For the parishioner, the institution becomes the insurance policy for assuring eternal life.

Cult of Personality

The history of religion is littered with personalities around which unhealthy attachments develop. In some cases, it may be someone in the capacity as the head of the church. In other cases, it may be someone held out as having a prophetic gift. Without critiquing individuals per se or the content of their message, it is not difficult to recognize when personality assumes too high a pedestal. Perhaps readers will remember two occasions in recent years—one event in the Netherlands, the other in Paris, France—when a cartoon about a prophet resulted in the death of many innocent people. Should not occurrences of this sort represent a classic case of idolatry of personality? Even if there is a perception that someone revered has been slighted, is there any context in which civilized people can give moral or ethical justifications for the slaughter of innocent lives?

When Tradition Trumps Reality

Finally, I would like to mention the propensity to allow tradition to assume an unhealthy level of authority. This is common among many ecclesiastical organizations where the tradition becomes more important than the reality. Once dogma is established, it tends to metastasize to the institution, and at that point data no longer matters. An example that comes to mind is quite personal, for in the faith tradition of my birth the dominant view has long been of reading the first few chapters of Genesis on very literal terms. The possibility of scientific or other evidence having anything to say seems out of the question, with many leading Church administrators holding this view with a great deal of rigidity. There are, of course, examples from many other faith traditions, where every possible rationalization will be put forward to defend an established tradition by minimize problematic data.

Conclusion

I recognize that some readers will probably have reached the conclusion by this point that the discussion in this chapter has been for the purpose of discrediting Scriptures and other categories of sacred authority, and to the extent that this is the case I would counter that the actual purpose of this discussion has been missed. The real objective has been to rebalance the knowledge equation. The simple truth is that *revealed authority* along with other sacred authorities outlined in this chapter do not come with a level of incontrovertibility superior to that of sense-based realities. This chapter has simply pushed back against this fallacy given its prevalence within the religious community.

I am mindful that many bring certitude of thinking to sacred constructs, where assumptions are predicated on an exacting correlation between mind and reality. In the real world, such correspondence is much more elusive, and particularly the ethereal realm given the nature of how sacred knowledge accrues.

This issue has been important to consider because over the centuries humans have often elevated sacred ideas well above an appropriate level, and this becomes painfully obvious when such ideas fail to mesh with verifiable facts. Such eagerness to overstate sacred claims must be viewed within the context of the human condition, with a longing to connect meaningfully and tangibly with the transcendent. Yet the urge to overstate the case through the levers of sacred authority must be regarded as pernicious because of its diminishing effect on the credibility of the entire sacred enterprise. Some wear contrarianism as a badge of honor, but unless it is backed up by a process that gives due respect to sensate reason, it is hard to argue that the sacred cause is being advanced in a way that will produce relevance.

In closing, this chapter has considered some of the reasons why sacred constructs have vulnerabilities, and there are several objectives that have been attempted as follows:

1. From this discussion the categories of the sacred have been considered, with reasons given as to why and how Scriptural authority is sometimes elevated way beyond an appropriate and intellectually defensible level.

2. Specific consideration has been given for reasons why the doctrine of inerrancy is fallacious.

3. Closely related to #2, consideration has been given to possible responses of seemingly conflicting passages.

4. Consideration has also been given to why, if Scripture were literally "God's inerrant word," would he communicate conflicting messages?

5. Consideration has also been given to the range of interpretations that are often represented.

6. Exposure has been made of the many scribal errors that crept into the oldest existing manuscripts—grammatical variations that number into the hundreds of thousands, and in some cases creating issues having doctrinal significance.

7. Note has been taken that none of the original manuscripts have survived, nor in many cases is it even known who the authors were. The implications for the claim that the books that make up the Bible were genuinely God's "inerrant word" spills over into a question of why, if God had intended humans to have his inerrant word, would he not have seen to it that the original manuscripts were preserved with no questions about variant versions—otherwise how is it possible to verify the original words, much less make a claim of inerrancy?

8. Finally, consideration has been given to a range of authorities that seem more designed to address human frailty than to represent justifiable constructs. A case can be made for understanding Scriptures to be a human document about God, rather than "God's inerrant word." This does not negate the prospects of inspiration in the process, but it does suggest cautionary wisdom as to the claims and assumptions made, particularly since areas of authority have been identified where it has been elevated way beyond all common sense and logical levels of appropriateness.

In connection with this last point, let me propose the importance of representing the "sacred" in a way that is not inconsistent with reality as being more urgent today than in former times. After all, humans live in an information age where it is much more difficult for a false construct to succeed. For those nurtured in a fundamentalist tradition where assumptions sometimes fail to match findings from the real world, an intellectual crisis is born when discovery is made that the actual reality does not live up to the assumed framing of that tradition.

In the next chapter consideration will be given to a number of other problems that emerge from a Judeo-Christian construct. Certainly anyone interested in developing a defensible worldview will need to understand where the vulnerabilities are, and it is in this context that reflection will be given to some of the vulnerabilities and dilemmas of faith.

CHAPTER 6

Dilemmas of Faith

Chapter Summary

The human tendency to impute more authority to a variety of sacred institutions than what rational scrutiny would deem appropriate was discussed in the last chapter. But such conclusions can be further reinforced by what, collectively, can be thought of as dilemmas of faith. In this chapter I consider four of these dilemmas. This does not mean that these are the only dilemmas, but surely they would rank among the more compelling. First, among these dilemmas is the issue of divine ambiguity, something that can be evidenced by the multitude of religious constructs that exist. In short, all understandings of religious truth are splintered and fractured, and this alone should propel modesty of claims made. Any attempt to approach this subject along objective lines will be fraught with problems, for in the final analysis, the ethereal realm is by nature ambiguous. Second, the problem of suffering antagonizes the concept of a loving God. A number of paradigms have emerged over time that attempt to resolve this issue, yet in the final analysis none of these paradigms is without problems, the details of which this chapter touches on. Third, many parts of Scripture would seem to suggest that God is a puppet master who controls the events on earth, yet as was discussed earlier, much that goes on can be explained by way of natural law—law that in many cases can be reduced to mathematics. Many times it even has a certain definitiveness attached as when a comet is on a path that will lead it to strike a planet. But there are also occurrences that have a less certain outcome, such as those based on probabilities, standard deviations and confidence intervals. In light of such realities, we are forced to contemplate such matters as creating a mystery as to the role God might play in the affairs of earth, since things do appear to pretty much run on their own terms. Fourth, somewhat related to the last category, for much of human history the behavior of the natural order lacked rhyme or reason, thus the ordinary tendency to appeal to the gods. However, with the rise of experimental science an increasing number of superstitious ideas gave way to empirical and rational expression. The challenge must be to ponder these dilemmas and recognize that not all theological problems can be resolved.

⸺

In the last chapter consideration was given to a number of ways in which false ideas form around certain sacred authorities, where they have been elevated beyond logical justification. The argument, there, urged a more realistic appraisal of all sacred authorities, thereby creating a more defensible framing.

In this chapter, the conversation will push forward into areas of religious dogma that are unaccountable to human sense and reason. For this discussion, perhaps it is best to start with a level playing field by noting that all worldviews come with certain loose ends that generally get tucked away out of view. The tacit element plays an important role, here, for obvious reasons. But to whatever extent the sciences deal with loose ends, sacred constructs must deal with their own set of vulnerabilities, exposing certain dilemmas of faith. By understanding such dilemmas, it will make it easier to rebalance the equation when conflicts arise between religious dogma and scientific conclusions. If nothing else, this exercise will reinforce the importance of humility in matters of an ethereal nature.

This particular discussion will examine four loose ends in Christian theology, three, of which, were outlined by two eminent Christian scholars, Philip Clayton and Steven Knapp, in their book, *The Predicament of Belief*.[1] These four areas include: the ambiguities associated with the nature of reality as evidenced by the great variety of religious pluralism—what some might regard as the opaque nature of the divine; a second line of reasoning has to do with the philosophical challenge regarding the problem of suffering—a subset of the broader category of the problem of evil; third, would be the fact that theological claims are sometimes incongruous with the mathematical realities; finally, theological constructs sometimes run contrary to certain scientific facts, leading to a worldview dilemma, particularly those who find value in a given religious narrative, yet also wish to live in a world that respects the senses. Each of these issues will be dealt with in the balance of this chapter.

1. See generally Philip Clayton & John Knapp. *The predicament of Belief*, Oxford University Press.

Ambiguity of the Divine

An acquaintance of mine who struggles with her understanding of God made the following statement recently—to paraphrase:

> God created, but then apparently chose to hide—emerging only periodically in ways that are subjective and not empirically verifiable. As a result, humans wishing to know him and his will are left to fill the silence with their own opinions and views. Some are subsequently certain God wants women to be ordained. Some are certain of the opposite. Some are certain God approves of them flying airplanes into buildings. Some think God doesn't approve of any killing, while others think God only approves of "their" killing. Fingers are pointed, and arguments ensue over what God really wants; heretics are cast away; churches splinter; new religions arise, and through it all we stumble around guessing.[2]

To be sure, the ambiguity of the divine is a fundamental issue for humans who grapple with the many loose ends for making sense of reality. Many people fail to give much attention to the issue of the hiddenness of God. However, some do count it as a mystery why a loving God who has care and concern for humans—their plight and suffering—would be so elusive, and why concrete knowledge of God is so impossible to nail down in any empirical sort of way. Christians often counter this mystery by pointing to the magnificent story of the Gospels as an example of God self-disclosing, yet when digging into the provenance of the record that has survived, there do remain a lot of questions and uncertainties, and this, itself, is a mystery. Why would God allow such a critical piece of evidence to become clouded?[3]

Indeed, why is the nature of the divine so vague?

In a global context, perhaps the clearest way in which to understand the issue of "divine ambiguity" is to imagine an unbiased individual who is on a sincere spiritual quest seeking enlightenment as to purpose and meaning of life—someone who is open to discovery of the divine and takes affirmative steps of discovery by embarking on an objective mission that seeks not to prematurely dismiss any particular path. Where should this search begin? Well, the obvious place to initiate this process would be to evaluate the many and

---

2. This paraphrased version that appeared in comments to an online article I published a while back. The commenter goes by the name of BethAgain, from the following article, http://spectrummagazine.org/review/2014/06/24/spiritual-brain-scientific-examination-religious-beliefs.

3. See note 40, for sourcing on problems with the New Testament's provenance, the many unknown authors, the variations in manuscripts, etc.

varied belief systems in existence, and to review the numerous sacred texts that might then lead to successful and rewarding discoveries.

There are perhaps 4,000-5,000 different religions in the world, with some of the best known being Baha'i, Buddhism, Christianity, Confucianism, Daoism, Hinduism, Islam, Jainism, Judaism, Shinto, and Zoroastrianism.

So where should this quest begin?

It must begin somewhere—perhaps with Christianity. Yet, anyone on the hunt for the most authentic Christianity will find this to be an intimidating task, given the large array of denominations. Here is the problem: with the estimated 41,000 Christian denominations in existence, it quickly becomes clear that interpreting Scripture is not all that easy, for if it were otherwise it could be assumed that there would be a very small number of denomination, or at the very least—one set (or a very small set) of doctrinal understandings.

Yet a thorough quest does not end here for, as noted above, there are many other religions in the world—many of which are also divided and fractured into a variety of subgroups much as Christianity is. Many also have their own sacred texts and prophetic traditions, claiming they are God's chosen people. In short, all understandings of religious truth are splintered and fractured.

By what criteria does a seeker begin this search, and how does an objective quest evaluate such a multiplicity of traditions—each vying to represent the face of God to the world? Secondly, why would a God of love leave humanity with such uncertainty? After all, this multiplicity issue exists because of the lack of clarity that humans have on matters of ultimate reality. This becomes obvious by simply referring back to the discussion in the previous chapter regarding some of the problems that exist with the Judeo-Christian Scriptures—the conflicting perspectives, as well as the state of historical evidence. While many denominations claim to be God's true church, basing belief upon a given understanding of Scripture, the simple problem is that there are many ways to interpret critical passages and this fact explains the multiplicity of denominations. The human problem, then, is summed up by the query of how a seeker should go about defining ambiguous passages that in some cases constitute fundamental doctrinal understandings? Every theistic religion on the face of the earth struggles under similar burdens to a greater or lesser extent, not to mention the problem of determining the criteria to be used to authenticate private testimony (or revelation)—something theistic religions are built on?

Many faith communities put on pretenses that the reality is other than this, and human nature tends to conveniently overlook these loose ends by choice or in some cases by ignorance. There is, in fact, a certain psychological power in communing with like-minded individuals, creating the sense that the reality articulated by way of narrative can be willed into existence if enough individuals are participating in its propagation. But ultimately, it is the ambiguity of history, as well as the ambiguity of the actual record that leads naturally to a plurality of religious traditions.

Some may argue that God evades direct perception because he wants to guard human freewill. The idea here would be that if humans had knowledge that was too absolute regarding God's existence and will for the world then human freedom would be compromised. Yet, there are no examples at the human level of too much information compromising the ability to clearly analyze and assess reality? Simply put, a commitment to theism must be based on the conclusion that God operates outside of direct sensate detection for reasons of his own choosing.

Analytical minds have no choice but to acknowledge the existential ambiguity of divinity, as well as the epistemological mystery it entails. At the end of the day, those who contemplate how knowledge is acquired (in the true "knowledge" sense of the word) must confront the fact that ethereal matters do not lend well to categorical thinking, and must therefore be seen as a dilemma for those who seek a more definitive understanding.

Problem of Suffering

The dominant Christian construct articulates the belief in a loving God who is omnipotent and omniscient (all-powerful and all-knowing). Yet, by following such a construct to its logical conclusion, a mystery emerges in light of the harsh realities of life and living. Is it possible for a loving creator of the universal order to be the author of death and destruction, or to stand by while suffering of sentient creatures abounds? This has, in fact, been one of the leading challenges the Christian faith has had to face. The late Charles Templeton gave voice to this problem in his book titled, *Farewell to God*.[4] In it, Templeton, a Princeton University educated Christian evangelist and pastor working with Billy Graham, later transitioned to agnosticism, and he addresses the problem directly. Primary among his reasons for doubt was the problem of the destructive forces of the natural order and its impact on life. He grappled with the same issues that many others have puzzled over, namely that in a world supposedly created by an orderly, loving, all-knowing, and all-powerful God, the world is full of diseases and natural disasters which are an inescapable part of reality. Why?

---

4. See generally Charles Templeton, *Farewell to God*.

He mentions some of the perennial questions:

- If God is love, why does he permit earthquakes, tsunamis, tornados, hurricanes, droughts, and other natural disasters to indiscriminately kill tens of thousands of innocent men, women and children?
- If God is a being of order and is creator of the universe, how should one account for randomness and disorder that adversely impacts sentient beings—frequently evident in nature via natural disaster?
- How could a loving God include as part of his creation such horrible illnesses as encephalitis, cerebral palsy, the various cancers, Alzheimer's and other incurable diseases, and permit them indiscriminately to afflict tens of thousands of men, women, and children?
- Why would a loving God have created parasites whose existence depends upon creating misery and death for sentient creatures?[5]

For sure, there is a lot of beauty in the universe, but what about some of the ugly parts? Is it possible that a loving God created with suffering as a central part of that paradigm? Claiming a role for God, however nebulous, has real implications.

One of the better-known, and succinct framings of the problem traces back to Epicurus who stated:

> Is God willing to prevent evil, but not able? Then he is not omnipotent
> Is he able but not willing? Then he is malevolent
> Is God both able and willing? Then whence cometh evil?
> Is he neither able nor willing? Then why call him God?

This is the issue known as theodicy, which deals with the tough question of how a loving God could coexist with pain and suffering. It comes with a lot of mystery, but offers no completely satisfying answer.

Central to this issue is the fact that the biological order itself is predicated on suffering with its predatory features, and while some of the harm that befalls humans is as a result of what the Christian world regards as a fallen moral order, not all suffering can be attributed to moral failure. The fact is the natural order plays a prominent role in this regard, it having nothing to do with moral failure. So, as thought is given to the path that Charles Templeton took, it does follow a certain logic, for as tragedy is observed daily from a ringside seat in front of the television news it often becomes

---

5. Ibid., 193.

the source of great perplexity, particularly in light of the claim that God is loving, and all-powerful.[6]

Traditionally there have been several ways in which to deal with this matter, and I will very briefly outline three of them below.

First and foremost, the predominant Christian view, sometimes known as the "divine sovereignty" or "blueprint" model (discussed previously) had roots in early Greek thinking, and later became integrated into Christian theology by St. Augustine, and after which was picked up by John Calvin. This general approach is so named because it views God as exercising complete and absolute sovereignty over the world and universe in all its minute detail. Consequently, all that happens, both the good and the bad, is deemed to be a part of a larger divine plan.[7]

Many who hold this view argue that whatever happens—for good or not—a loving God uses to achieve a higher good purpose. Because of the way these understandings are framed, the problem of tragedy has traditionally been one of formulating a loving and good purpose behind tragic events—i.e., what humans perceive as evil, bad, or tragic, actually serves a specific divine purpose. Consequently, the human problem is one of limited perspective, with an inability to see the grand scope of God's overarching plan.

Yet the dilemma, as noted above, in elevating God's sovereignty in this way is that it logically implicates him for everything that would be characterize as wrong with the world, and this reality forces the exponents of this view to face its main problem—that is, harmonizing a good God with a world that is very flawed.[8]

For sure, there are some who find comfort in the notion that an episode of tragedy, death, or destruction is part of God's plan and that only if finite humans could see the larger picture it would all be acceptable. But this attempt to exalt God by asserting his sovereignty has a troubling outcome,

---

6. The idea here is that there is little to recommend the path of faith unless it has rationality behind it. While there are a number of variant constructs for handling the problem of theodicy, sometimes the constructs create new problems. Also I should mention that one of the most readable books regarding the problem of theodicy for which I am familiar is the book by Bart Ehrman, titled *God's Problem*.

7. Augustine, *Confessions*, trans. H. Chadwick, 125; see also Gregory A. Boyd, *God at War*, 45. Here Boyd points out that Augustine explicitly held that for God, evil does not exist. While this specific page is cited, Boyd discusses the Augustinian point of view on this matter rather extensively throughout his referenced book. Essentially, to make the Augustinian argument at the divine level, we must assume that the end does justify the means! As an alternative view, Richard Rice, Clark H. Pinnock, John Sanders, William Hasker and David Basinger, in, *The Openness of God: A Biblical Challenge to the Traditional Understanding of God*, present other possible ways of formulating theism.

8. Ibid as to Boyd, *God at War*, an evangelical scholar, who develops a cosmic conflict theme.

in as much as it becomes the context by which he ends up responsible for all of the suffering that takes place in the world. Certainly if God's exhaustless sovereignty is to be fully embraced when faced with a natural disaster of any size there is a basis for calling God's goodness into question, and the issue is even more acute in the face of a disaster on the scale of the 2004 Indian Ocean tsunami that killed well over 200,000 people. Surely a sovereign God who controls all things, and who is by definition, good, would not allow such catastrophes to occur. This, at least, is the conclusion of Charles Templeton and many others who have demonstrated the deficit in such a formulation.

A secondary approach is somewhat of a subset of the "blueprint" model, in as much as it also embraces divine sovereignty. Because of this fact, not a lot will be said about it, but it does deserve mention in as much as it does attempt to propose a concrete way by which to navigate through the problem of theodicy. It was first proposed by Irenaeus, a second century philosopher and theologian, and has been embraced in modern times by theologian John Hicks who championed it in his book *Evil and the God of Love*. The general idea here is that this world was created to develop moral characters and to assist humans in drawing closer to God—what John Hicks refers to as "soul making." Yet anyone who contemplates the magnitude of suffering along with the vast imbalance of it in the world must surely have doubts. Most readers will know of individuals who have suffered vastly more than could be represented in any balanced, just God.[9]

As a result of apparent shortcomings that flow from a sovereign God model some have found motivation to pursue a different explanatory path—a third approach that I will label as the *cosmic conflict motif*.[10] It overcomes some of the more objectionable features of the sovereignty model by introducing a way of removing from God the responsibility for all that goes on, on earth, arguing that there are forces outside of God that have compromised the perfect order he created. In fact, Scripture does ostensibly give voice to the idea of *The*

---

9. For a related discussion, see John Hicks, *Evil and the God of Love*, 43-95; later published by Harper & Row: San Francisco (1978). It should be noted that the prominent theologian, John Hicks, has adopted the Irenaean/Augustinian approach by suggesting that evil plays an important role in the world in fulfilling God's purposes. He proposes the possibility that God envisioned evil from the beginning, and that it contributes to the process of "soul-making" by enabling humans to develop positive traits of character. Hicks suggest this approach because he has difficulty with the idea that perfectly created intelligent beings would, without cause, rebel against God.

10. This approach also goes by a variety of other framings. Ellen White, a nineteenth century author popularized the term, *the great controversy,* and authored a book of the same name though had no theological training; Gregory Boyd, referenced above, terms it as *warfare theology*; others have proposed a variety of other terms, but the common thread requires rejection of the divine sovereignty model.

Fall, where a perfect beings endowed with moral freedom, took that freedom in a direction that departed from the laws that govern the created order. Such thinking has the capacity to offer logic to suffering caused by moral failure in a cause-and-effect world created by a loving God.

This overall approach traces back to the sometimes apocalyptic views expressed in the New Testament, and is perhaps best known from the literary works of John Milton, who sort of developed the framework of this approach in his classic, *Paradise Lost*. More recently Gregory Boyd has developed this in a more systematic way through his writings. Significant to this view is the idea that God is not fully sovereign and thus many things that happen are not attributable to him. The logic for this view argues that God created free moral beings, and in the face of genuine freedom the integrity of the divine order was contingent upon moral beings living within the framework of the divine order. As this narrative goes, the divine order started out as perfect, with created beings endowed with moral freewill. But this order was compromised when discerning moral creatures opted for a different course. In a cause and effect universe, this had real consequences, and as a result the moral order now operates outside of its created design, with humanity now suffering the consequences. Furthermore, in a free moral universe, the divine order can ultimately only prevail if God allows the alternative order to play itself out so that all free moral beings will have a graphic demonstration of the consequences of acting outside the parameters for which the universe was constructed. Most who adhere to this approach believe that in the fullness of time, God will rectify this problem and will set the universe back on its designed course—but only after the full consequences of operating outside the divine order are fully mature.

If I were to succinctly distill this approach down it would include the following points related to a cosmic conflict ongoing in the universe that has compromised the world and/or life in it in certain ways:

1. It begins with a theistic premise, and if this premise is informed by the Gospel record, then incorporated into it would be the idea of a loving God who created a moral order;

2. That if God created a moral order, then both good and the potential of the inverse of the good was built into the structure of reality, with the quest being one of distinguishing between the two;

3. That if God is good, then evil must be understood as a force in conflict with the government of God;

4. That if evil is a reality, then any thoughts of a good God having exhaustive and complete sovereignty of the universal order needs to be

qualified. The idea here is that in between God's sovereignty and human reality, is a force of some cosmic significance that is at war with God and his established order—thereby qualifying any discussion of God's sovereignty. It is the influence of this contravening force that can provide an explanation of the processes that pertain to many aspects of tragedy, and do so without necessitating any need of viewing such occasions to be some part of a divine plan.

An excellent example of the cosmic conflict theme in the New Testament can be found in the Book of Mark, who prominently places the events surrounding Jesus within the context of the struggle between good and evil in the universe. In *The Origin of Satan*, Elaine Pagel, whose study on "Satan" treats him as a mythical literary figure, notes that in the context of Mark's narrative, Satan and his forces retaliate against God by working to destroy Jesus beginning in chapter 1. Matthew and Luke, authoring their Gospels several years later expanded upon Mark's opening presentation. Pagel notes that all four Gospels, with variation, depict Jesus' execution as "the culmination of the struggle between good and evil—between God and Satan—that began at his baptism." Pagel further notes that with Mark's opening narrative of Jesus' baptism, "God's power descends and at that moment humans disappear from the narrative and the spirit of God drives Jesus into the wilderness to encounter Satan." From this account, the inference to be drawn is that these events convey that struggle between good and evil in the universe, which the narrative itself suggests is something beyond a mythical reality.[11]

One of the strengths of this approach has to do with its internal logic, making intelligible a way of thinking about how a good God could coexist with the human reality of pain, suffering, and death. It can also provide an explanation for the necessity of the Christian concept of the Incarnation where estrangement and fracture have reduced the human capacity to comprehend the divine nature and will—that in a moral universe where moral beings have a choice between good and evil, it becomes necessary for God to reveal himself in a personal way if he is going to be able to restore a sense of the divine order among moral beings.

---

11. On this point, it should be noted that while the Old Testament does not define "Satan" with the same clarity as that of the New Testament, nevertheless, there are many Old Testament passages that do suggest the cosmic nature of evil in the universe. It would also be appropriate to mention at this juncture that attempts to propose a cosmic conflict theodicy outside of the Judeo-Christian tradition, goes back centuries in time. It was the sixth century BCE Persian teacher, Zoroaster, from the era 600 BCE, who proposed a metaphysical dualism between good and evil, resulting in the notion of competing deities. Even to this day, there are a few formal adherents to this belief system.

But as helpful as this overall approach might be, it also comes with problems. It is one thing to accept suffering as a result of moral failure, where, for example, an individual becomes the victim of violence, and quite another where suffering is as a result of the natural order. It is significantly more difficult to explain death at the hands of the natural order—an order that most Christian narratives identify as being authored by a loving God who looked at it and saw that it was good. It is the many events of the natural world that are destructive to life to which evangelical theologian, Norman Geisler, refers to as *physical evil*.[12] This terminology is consistent with some interpretations of Scripture that depict "the fall from grace" to be connected to the adverse impact the natural order is capable of delivering. In support of this view, in Genesis is found the narrative of the fall of humans from grace, and according to the record this occurring sometime shortly after creation—certainly within the lifetime of the first parents and prior to them having children. The Apostle Paul suggests that death is as a result of sin; thus, the implication that death did not occur before sin. This, however, would seem to run contrary to the evidence of the geological column specifically, which manifests a biological world where predatory death and tragic events of the natural order took place well before the appearance of humans.

The Fall cannot be validated empirically, yet there are ways to empirically test certain aspects of this traditional belief.[13] This is possible because of the theological tendency to limit disordering events of the natural order to a "sinful" planet earth. For this, there is the laboratory of the universe to evaluate the viability of these traditional ideas. From here, it gets sort of complicated, because for one thing cosmology has established quite clearly that the universe is a violent and dangerous place, with some of the natural hazards on earth also found in other worlds apart from earth. For one thing, the laws of physics here on earth are the very same laws that govern the entire observable universe. Many of the destructive hazards of the natural order seems to have a ubiquitous presence throughout the universe—places far removed from "sinful" earth. This raises fundamental questions about the logical relationship between natural disasters on earth, and human sin.

---

12. Geisler, *Baker Encyclopedia of Christian Apologetics*, 222. Geisler's use the term "physical evil" is also sometimes known as "natural evil." In discussing this, he has in mind natural disasters of all kind—earthquakes, tornadoes, hurricanes, etc. that Christians assume came on the scene after the fall.

13. From the perspective of physics, events of the natural order that produce disorder from the human frame of reference paradoxically have an underlying order—processes with the capacity to be explained in rational terms. An asteroid striking the earth can be explained rationally and thus there is an order about it, but the event itself can create a lot of chaos and disorder in terms of disruption to life.

As is common knowledge, death can come from a number of sources including, old age, violence, accident, and the many events of nature—tornadoes, hurricanes, earthquakes, tsunamis, floods, exposure to temperature extremes, and disease among others. Obviously moral dualism may be a way of explaining predatory behavior—over against the divine ethic—but it certainly does little to explain a destructive natural order that delivers pain and suffering, particularly given the narrative of a loving creator God. In fact, it is possible to isolate specific findings of science that would seem to contradict any attempt to tie the destructive features of the natural order to sin.

The reality is, there are many natural events originating on earth that can disrupt life. People of faith often identify frightful and destructive events of the natural order and ascribe it to be sourced in *The Fall*.

So the question emerges, "Are these natural events a function of *The Fall*?"

Quite a few individuals have suspect that they are; yet it is now known that many of these events are not unique to earth. When observation is made of other planetary systems and/or their satellites, evidence emerges of volcanic and seismic activity, of vast and fierce storms that are in fact much more severe than anything experienced on earth. The undeniable conclusion is that these life threatening natural events on earth also occur on other planetary systems—systems far removed from *sinful earth*. All of these events are, of course, sources of death on earth, and so when these unfortunate events occur, there are reasons to doubt that they could be legitimately viewed as a part of the sin problem, given their seeming ubiquity around the universe.

Another source of danger to earthlings are cosmic objects such as meteors, asteroids and comets. There is evidence that earth has been the recipient of such objects on many occasions creating widespread devastation to life. Occasionally, earth's gravitational pull captures these intrepid wanderers. In fact, most humans have witnessed meteor events, and even though in most cases they have little or no destructive impact, they occasionally have been documented to accrue significant damage. Asteroid events, on the other hand are by definition large and destructive, and while none have occurred in modern times there is significant evidence to indicate that they have visited earth in the past and that such events have been destructive of life. Earth is also believed to have captured comets at points in the past, and would have had the similar destructive capacity of an asteroid. Most readers will likely be familiar with the comet Shoemaker-Levy 9's collision with Jupiter in July of 1994, this being the first direct observation of an extraterrestrial object colliding with a planet in the solar system.[14] This observation is consistent with scientific findings that would suggest Earth has likewise been the recipient of these wandering cosmic objects.

Now add to this the fact that galaxies sometimes collide with each other, creating the potential of a lot of disruption, not only to a star and its satellite systems but also to any life that may exist near those events.[15]

All of this underscores the fact that natural processes occurring on earth are not hazards unique to earth, and therefore appears to be independent of the moral status of earth. The brutality of the material features of the universe towards biology appears from the human vantage point to be universal. Those who hold a belief in God are simply forced to grapple with issues that would seem to interfere with the common assumptions that the natural order has been corrupted by the taint of sin that has given it the capacity to neutralize life. If it were otherwise, there would be some expectation that the natural order would be observed as differentiated in other parts of the universe from that of sinful earth, yet this is not the case. Adding to the troubles for the cosmic conflict perspective is the evidence that comes from geology, where there is an orderly record of what has transpired on earth over a very long period of time. When the geological column is examined there is a generally consistent finding of an orderly progression of life, begins at the lower levels of simple life forms, and from this simplicity there is a progression of complexity up the column that cannot be explained by a universal flood or any other conceivable biblical narrative. This suggests that biology on earth existed in varying forms long before humans entered the scene, and thus that biological death existed on earth before humans could have sinned. On the face of it, this can be seen as an anomaly between the biblical narrative and the scientific reality unless some way can be found to find convergence.

In the final analysis, if the explanation of the human condition is tied to a cosmic conflict perspective it is natural to wonder how many millions of years (or even thousands of years for those holding to a parochial narrative) of suffering and tragedy of sentient creatures are required in order for intelligent beings to draw up mature conclusions about the choice of moral orders. Such speculation assumes there to be questions as to the fundamental nature of a moral universe that moral creatures must come to some resolution about—whether it works best when represented in the survival values

14. http://en.wikipedia.org/wiki/Comet_Shoemaker%E2%80%93Levy_9.

15 See for example, http://www.haydenplanetarium.org/resources/ava/galaxies/G0601andmilwy; see also http://blogs.villagevoice.com/runninscared/2011/08/collision_of_ga.php; see also http://cas.sdss.org/dr3/en/proj/basic/galaxies/collisions.asp.

of the jungle, or alternatively oriented outward in kenotic self-emptying as expressed by the divine ethic?

In concluding this section, then, it seems that all three paradigms of suffering discussed above come up short. At this stage, there simply are no formulations that are problem free, and in this respect, it extends the dilemma discussed above of God's hiddenness, for it seems that humans have not been able to come up with a way to understand a God of love in the face of the anguish and sorrow that humans all experience. Unfortunately, this remains as an issue unresolved.

The Mathematical Challenge to the Faith Narrative

In some respects, this section overlaps the prior section dealing with human suffering. However, it is distinguishable in that the approach is on mathematical terms. In this regard, I would like readers to recall the discussion back in chapter 4, which discussed some of the issues raised by Leonard Mlodinow. It may be recalled that he traced the development of mathematical theory of probability, pointing out that while the ancient Greeks were instrumental in moving mathematics forward in a number of ways having to do with axioms, proofs, and theorems, yet they had no theories dealing with probability. He suggests that one of the reasons for this may of had to do with their general belief that the future in all its detail (even wagers) unfolded according to the will of the gods, thus leaving no place for a theory of randomness and probability. By subscribing to this deterministic worldview, the Greeks seemed to have overlooked the possibility that events might occur on a random basis.[16]

The documented history of humanity is one dominated by superstition—particularly events involving the natural order. Not only the early Greek cultures attributed much of what transpired within the natural order, to God, but also many earlier civilizations, including the early Hebrews, as well as the polytheistic Mesopotamian cultures. When events turned destructive, or even deadly, there was generally a divine explanation created to account for what happened. For the Hebrews, the unfortunate happenings signified God's anger over estrangement from, or disobedience to Him—often understood as a community that was not sufficiently tuned to the divine will.

Thus, the idea of random processes existing upon earth is a radical departure from this earlier understanding and has enormous theological implications. This is particularly the case for those nurtured on Augustinian thought, for the existence of randomness implies that "mathematical

16. See generally Leonard Mlodinow, *The Drunkard's Walk*.

probability" plays a role in the processes of the natural order and in human experience, and that aspects of the future may therefore be open and undetermined in the present.[17]

The outcomes that random factors produce can frequently be calculated with a mathematical degree of confidence and it is on such basis, for example, that one can determine the probability of how many will die from traffic accidents along a time/location coordinate. A determination such as this can be made from an accounting of the variables and the statistics of previous occurrences. Some of the variables might include things like the volume of traffic, time of day or night, weather conditions, vehicle condition, driver attentiveness, etc. The biggest variable of all would be the victims. It is not possible to know who these victims will be ahead of time. All that can be known is the probable outcome. To the extent that the significant variables can be accounted for, it is possible to make reasonably reliable predictions. Those who travel a given route either once or on a regular basis now become factors (or data elements) in this probability distribution where unplanned events occur (including events of a tragic nature). When the worst possible outcome occurs (tragedy), it is possible to conceptualize other possible outcomes if the participants to the event had altered their schedules by only a few seconds. The point is that being a factor does not destine a person for a given fate, but random events can sometimes align to make these tragic outcomes possible. This is what makes life so unpredictable.

Likewise, epidemiological probabilities of acquiring a specified disease given certain relevant variables—including genetic predisposition, lifestyle, geography, etc.; or the sociological probability of becoming a crime statistic—age, race, location, time, etc. In all such scenarios, if the samples are drawn properly and an adequate accounting for the variables has been made, it is possible to make such determinations with a mathematical level of confidence.

As a part of a discussion of randomness there is another somewhat related issue that comes out of science and that pertains to the fact that very

---

17. We can give brief consideration to mathematical probability through a simple example of a coin toss. We know that all things being equal if we toss a coin we have an equal chance that it will come up heads or tails. We also know that if we toss a coin 100 times, we will end up with roughly ½ of the tosses being heads and the other ½ being tails. However, if we toss a coin once and get heads, we cannot assume that if we toss the coin again it will be tails. Nor can we assume that we have an increased probability of a tails on a second toss. That is because each toss is an isolated event unrelated to any prior toss, and each toss will have a random outcome. Thus, the probability remains 50/50 that we will get a heads on the second toss, and the probability of each subsequent toss remains the same regardless of what has transpired previously. In other words, each toss event is unique and unpredictable, and free of any influence from what has transpired previously. Ultimately all that we can do is work with the law of averages based on multiple events, and from it develop a probability distribution.

small and seemingly insignificant factors related to an initial process can have enormous later consequences in terms of how a process unfolds. As a point of illustration, whether or not an incident occurs, or whether an incident becomes a tragedy may involve nothing more than a seemingly insignificant change in schedule involving just a few seconds. Such seeming insignificance has the capacity to completely rewrite history, including the possibility of whether or not future progeny will have the opportunity to materialize.

It is possible to think of decisions made in the remote past that did not necessarily seem monumental at the time, but in retrospect can be seen as having been pivotal and life-changing, profoundly influencing outcomes. It could be something on the order of a chance meeting of an individual, who turned out to become a life-mate, or a future employer. To take this a step further, the miracle of each individual is that everyone alive can track their origins back to an exact alignment of individuals (a mother and a father), sperm and egg, out of trillions upon trillions of alternatives (where people-pairing possibilities and coitus timing and conditions are involved). The odds become even more exponentially incredible by moving back beyond the current generation to ancestors where a perfect alignment of all these variables are required in order for each individual to materialize. This is probably the most compelling human example of random factors shaping reality, and how seemingly insignificant elements can ultimately become very significant.

Those who have been nurtured on the idea of divine control of all that happens may have difficulty in cognitively processing or acknowledging the power of mathematics and the reality it describes. In facing the statistical data, including a study of the victims of tragedies, good reasons can be found to believe that God-fearing individuals are not excluded from the laws of probability and random processes unlike their secular counterparts. Bad things do indeed happen to good people, and the story that mathematics would seem to tell is that there is little room for reading any theological meaning into tragedy—in terms of divine will. Such events are simply a fact of life and living, and must be explained in a way that does not create divine culpability if the concept of a good and benevolent God is to be taken seriously.

So once again those who view the world functioning along deterministic lines controlled by the divine will, are left with a dilemma in that such events may fail to line up with the indeterminate reality as it is found in nature beginning at the quantum level. Mathematics presents a rather unassailable portrait of the world as it is. It tells of a world that is not the deterministic product of sovereign action, but rather one that is governed by certain random processes that can sometimes create adverse outcomes—including suffering, and death. Although the mathematical critique may at first blush seem to elevate the cosmic conflict approach, it does little to overcome the dilemma previously detailed of the universality of natural random processes that have the potential for being destructive of life. Whether the universe is understood along determinate or indeterminate lines problems still exist, even if the cosmic conflict approach seems more charitable towards deity.

The Scientific Challenge to the Faith Narrative

As previously noted, for most of human history the behavior of the natural order lacked rhyme or reason, thus the ordinary tendency to appeal to the gods—Mesopotamian, Hebrew, Greek, or otherwise. However, with the rise of experimental science an increasing number of these superstitious ideas gave way to empirical and rational expression. The success of science in logically explaining mysterious phenomena on strictly natural terms—with no need to invoke supernatural causes—has for many people led to a diminished role for the divine. The effect of all these discoveries has been that, as Clayton and Knapp put it, "the universe increasingly looks like a completely closed and self-explanatory system."

On top of that, there are a number of traditional interpretations of Scripture around which science has something to say, and in some cases the traditional interpretation is at variance with science. For example, the Genesis account of beginnings provides a description of a progressive creation that begins with simple life forms that then moves to increasingly more complex forms, and while this is consistent with scientific findings there are other parts of these two narratives that are viewed by some to be in conflict. The data from the geological column tends to validate a progressive aspect to biology, but the rub comes with the evidence that such progression extended over a vast period of time. For some, any notion of deep time is a major problem since the biblical narrative speaks of the creation in terms of a succession of days over the period of one week. Adding to this conflict was the Apostle Paul view that death was caused by sin, yet if life evolved over an extended period of time—with the life and death cycle occurring well in advance of human life—then there exists an apparent conflict with this Pauline statement given the data suggesting quite unambiguously that the life and death cycle existed long before humans appeared on the scene where the question of sin could have been a factor in death. There may be creative ways to handle this dilemma, but not if the biblical account is to be read literally.

Then, of course, on top of this there is a great deal of evidence that would support an evolutionary process in the development of biology life. But if the starting point is of a loving God, then the fundamentalist objections are not entirely irrational since it does seem problematic that the divine creation of life would be based along predatory lines where survival is supported by fitness—not to mention the hazards of the natural order that are sometimes unrelated to questions of fitness.

Thus, the findings of science must join the other dilemmas about which there really are no satisfactory answers. If the person of faith can proceed with flexibility, the potential grow for interpretations that can account for such apparent inconsistencies. But flexibility must also include a large degree of humility about what can be known through revelation, and the degree to which expressions of certitude are appropriate.

Conclusion

To summarize this chapter, then, there exist multiple dilemmas for those who either seek faith or are immersed in a belief system that offers some ultimate meaning. The dilemmas range from the ambiguity of the divine, to the problem of suffering, the incongruities of mathematics with certain constructs of faith, and finally scientific data that put the challenge to certain interpretations of Scripture. These are each, dilemmas that must be faced and to which there is no adequate answers—each having the capacity to challenge the faith narrative in certain particulars.

Thus, this chapter winds down without offering up satisfying responses. The challenge must be to ponder the paths that others have charted, or to ponder paths yet to be dreamed in building a paradigm that can satisfactorily address these dilemmas. These are difficult issues to navigate, and the currents are not all that friendly. It seems that ultimately there is no way forward that is problem free, so these issues will influence that which may ultimately develop from such reflection. As beings born to optimism and hope, the quest for answers must go on.

In the next chapter, the focus shifts from external factors that impact theology to the internal dynamics of human physiology. As it turns out, the connection between the spiritual world and human physiology may be more important that previously considered. Research that has taken place over the past century has now settled on the conclusion that the temporal lobe region of the brain plays a central role in human spirituality, and some of the findings in this area of research will now be examined.

CHAPTER 7

The Spiritual Brain

Chapter Summary

The focus up to this point has been mainly on the process of knowing, along with the degree to which certitudes of thinking about the sacred are appropriate. However, there are possible physiological facts that deserve consideration as well. In a provocatively titled book, *Did Man Create God?* David Comings, who has formal training in neuroscience explores the physiology of the sacred, centered in the brain's temporal lobe. Approaching this subject on strictly naturalistic and scientific terms his answer to the question posed would seem to be that humans have created God. He notes the occurrence of very spiritual experiences under a wide variety of circumstances, including pathologies associated with epilepsy, oxygen deprivation, near death experiences, drug induced, and natural spiritual encounters. All of these conditions produce remarkably similar spiritual manifestations that can be measured in the science laboratory. It is his conclusion that much of religious history can be explained on this basis, and while it can be anticipated that some will defensively push back against such findings—perceiving it to be an attack on religion—it is important to recognize that in reality none of this can address the question of God's existence as that is a completely separate question. Yet, what it can do is provide a rational basis for understanding some of the history of religion, including perhaps some of the difficulties that can be encountered with sacred texts, such as has already been discussed.

Up to this point consideration has focused mainly on the process of knowing, along with the degree to which certitudes of thinking are appropriate. As it turns out sacred knowledge is in a category that is distinct from sensate knowing. Most thoughtful people of faith do recognize this reality on some level, though in my experience there are many who fail to see the nature of difficulties associated with sacred knowledge. It was for this reason that a chapter was devoted to some of the idolatrous ways in which humans seize on issues of ultimate concern. As if this situation wasn't complex enough, consideration was then shifted in the last chapter to areas of faith that are not all that tidy—areas where inconsistencies and incomplete information can leave room for doubt—exposing the human predicament in connection with sacred formulations. The conclusion to be drawn up to this point would be that categorical thinking on sacred matters does not mesh well with human finitude.

While such existential realities should inform how the framework of a belief system is constructed, in the real world it is quite easy to see that most people seem comfortable in proceeding blindly with certitudes of mind on ethereal matters. Such less defensible attitudes may on the surface seem puzzling, yet it seems relatively easy to explain as a part of human coping in an uncertain world. Convictions of mind create a more controlled environment for dealing with life's vicissitudes, but there may actually be a physiological explanation for some of this as well.

Moving now to the topic of this chapter, the point that I propose to be worth considering is that in spite of all these obstacles to sacred knowledge it is significant to note the pervasive universality of human concern for sacred forms of mythos, and that such universality can be explained in part by physiological factors as part of a natural human process. With this in mind, this chapter will consider briefly the possible role that biology itself might play in this whole equation.

In a provocatively titled book, *Did Man Create God?* David Comings, a physician, with formal training in neuroscience and genetics, offers up an underlying thesis that humans have created God. He supports this idea in a number of ways, but among these he explores with readers some of the latest brain science made possible with modern technology that came on line within the past few decades. It includes a discussion of the various control centers in the brain that have been identified, allowing for the ability to enjoy pleasure, engage in social interactions, execute rational thought, and experience the spiritual and transcendental.

The idea that sacred concepts are associated with the human brain physiology is likely not something that most give a great deal of thought to. But I would submit that while Comings approach is strictly naturalistic and has reached certain conclusions that exclude God, some of the ideas expressed in his thesis are not terribly controversial with a good many theologians who grapple with issues that sometimes emerge, including the different representations of God found in Scriptures and other sacred writings. It suggests a human component, involving human based projections to some extent. For sure, to reach such conclusions will have no necessary relationship to the question of God's existence, for it is logically possible that humans have created God in certain particulars, and that God also exists—it is possible that they both can be true.

What research has discovered more generally is that distinct and identifiable parts of the brain control for a variety of functions, including pleasure, social interactions, rational cognition, and the spiritual element within humans that is universal. In the following pages, consideration will be given to the relationship of brain chemistry to human perception of the world in all its permutations, including questions of the sacred.

Starting with the parts of the brain that produce pleasure, it is worth noting that there exists a dopamine-rich series of pathways that are responsible for gratifying effects of eating, sex, and love. These specific activities are necessary for the survival of the species and are called *natural rewards*. However, there are also a number of chemical substances (drugs) that can induce a synthetic pleasure, including alcohol, PCP, amphetamines, morphine, cocaine, tobacco, or other substances. Pleasure can also come from a number of non-chemical activities such as extreme sports, gambling, etc. As Comings notes, individuals with a genetic defect in their dopamine reward pathways are particularly susceptible to these addictions.[1]

Meanwhile, the social brain is engaged in people-to-people relationships involving parents and child, spouses or partners, work associates, and those relationships between friends and other. These parts include:

1. The limbic system, which adds emotional flavoring to life.
2. The amygdala, which teaches fear.
3. The hormones oxytocin and vasopressin, which are involved in pair bonding, monogamy, love, maternal care, and trust of others.
4. The orbitofrontal prefrontal lobes, which allow connectedness to emotions through the limbic system and the autonomic nervous system. This system controls emotional behavior and allows for empathy and understanding of the emotions of others. Comings further notes that antisocial behavior is common when this area is dysfunctional.[2]

---

1. Comings, *Did Man Create God?*, 311.
2. Ibid., 315–330.

The rational brain, or the thinking brain is controlled by the dorsolateral prefrontal cortex, and is the neural site of working memory, abstract thought, interactive decision-making, deductive and inductive reasoning.[3]

Finally, the spiritual brain is centered in the temporal lobe and is the central location that connects the individual to spirituality, and the evidence for this comes on many fronts. These findings further underscore the fact that the more spectacular manifestations of the spiritual sometimes involving some sort of stimuli, or in extreme cases, pathology.

Such findings include studies involving electrical stimulation of the brain in conscious subjects, spontaneous and drug induced hallucinations, temporal lobe epilepsy (TLE), near death experiences (NDE), oxygen deprived brains, changes in cerebral blood flow, stress, music, traumatic lesions of the temporal lobe, and studies of individuals who manifest a range of spiritual emotions, including meditation and the ecstatic charismatic states associated with Pentecostalism. It is particularly noteworthy that the pathology associated with TLE, which is traceable to a variety of causal factors, including traumatic brain injury, is particularly vivid in its impact on the spiritual realm. As Comings puts it, " . . . over a period of many years it became apparent that there is something uniquely spiritual about seizure activity that emanated from the temporal lobes."[4]

Temporal Lobe Epilepsy (TLE)

TLE was first documented more than a century ago, but received its largest initial boost from the clinical work of Wilder Penfield back in the first half of the twentieth century, with him documenting its relationship to profound spiritual encounters.[5] His mapping studies place both the right and left temporal lobes as the center of a variety of complex spiritual experiences that come in a range of manifestations, including trances, out-of-body experiences, automatic behaviors, feelings of being in the presence of others,

---

3. Ibid., 315.

4. Ibid., 347, and 392; it should also be pointed out that Comings is not an isolated voice on this subject. He, in fact, has provided a wealth of documentation within the sciences for these insights. Another helpful book in this topical area is "Sword and Seizure: Muhammad's Epilepsy & Creation of Islam," Abbas Sadeghian PhD.

5. Penfield, W. The role of the temporal cortex in certain psychical phenomena. *The Journal of Mental Science.* P. 424; Penfield, W. the permanent record of the stream of consciousness, *Acta Psychol* (Amst), 11; p. 47–69; Penfield, W. & Perot. The Brains' record of auditory and visual experience. A final summary and discussion, Brain, 86; 595–696. These are all relatively old sources, but is recognized and cited approvingly by David Comings as referenced in this chapter.

angelic voices, of "intense meaningfulness," of being "connected to some force greater than the self," and/or direct interaction with God.[6]

As noted above, Comings reports that one of the most fascinating features of TLE is the nature of the religious feelings that may occur, with it many times resulting in profound personality changes. Some of these changes result in religious conversions, even going so far as to effect change from one religion to another. One form of TLE seizure is so dramatic that it has been labeled "ecstatic seizure."[7] This was first documented and published in 1980 by EEG, and appears in two forms. In the first of these forms the patient is described as having an "indescribable sense of joy' and has been clinically identified as emanating from the right temporal lobe.[8] The second form is primarily "a cognitive experience of insight into the unity, harmony, joy, and divinity of all reality, usually with pleasurable effects.[9] Comings notes that a number of investigators who have studied ecstatic seizures associated with TLE suspect that it has had a significant impact upon cultural and religious history.[10]

A distinct and rare form of TLE is referred to as "the 4-H Syndrome— Hyper-religiousity, Hyposexuality, Humorlessness, and Hypergraphia. Essentially, this is a condition that is marked by a high level of religious concern, a muted sexual interests, takes life very seriously, and has a compulsion to write.

Near Death Experiences (NDE)

There have been many books written about NDE and the reports of those who have had these close encounters with death. Some of the features of this condition as reported are feelings of serenity, of being in a tunnel, often ending with a very bright light, and an out-of-body experience. For many NDE has been a meaningful spiritual experience, and proof of the existence of God, and of life after death. On this, Comings has what he calls "a more sober and scientific account" which has been documented in a very controlled clinical setting. This developed through recognition of something similar experienced by pilots who are subject to extreme G-force, where they would sometimes black out and have a hallucinogenic experience.

6. Comings, 254.

7. See Cirignotta, F., Tudesco, C. V. P & Lugaresei, E. Temporal lobe epilepsy with ecstatic seizures (so called Dostoevsky's epilepsy). Epilepsia, 21; 705–710, see also Comings, 356.

8. Ibid.

9. Ibid.

10. Ibid.

In order to study this condition, a large centrifuge was built in order to create similar G-force conditions in a controlled environment.[11] The discoveries that came out of this research seem to indicate that G-force can deprive the brain of oxygen, and as the brain becomes oxygen starved certain manifestations of mind occur and is likely caused by the release of certain neurotransmitters. As part of a chain reaction, this situation prompts the release of other identified compounds that can produce the hallucinogenic manifestation that is often associated with NDE. In the words of Comings, "rather than proving that God exists, NDE proves that when the brain is deprived of oxygen for prolonged periods of time, immediately prior to brain damage a range of physiological events occur that characterize NDE." He concludes from the research that those who suffer depression or fear death and then have an NDE, experience an overall improvement in mental function with greater stress tolerance and an enhanced appreciation for the spiritual aspects of life.[12]

Drug Induced Spirituality

Dr. Rich Strassman, a psychiatrist with the University of New Mexico, conducted a study of the effects of di-methyl-trypamine (DMT) in humans. The molecular structure of this drug is very similar to the neurotransmitter serotonin. From this study he has labeled DMT as *The Spirit Molecule*, doing so based on the effect it had on human subjects.[13] The effect of DMT can be very euphoric and pleasurable, and provided a powerful sense of spiritual connectedness, including interaction with non-human beings. One investigator reported that one of the most remarkable aspects of the DMT experience was the reporting of how real the experience of having had contact with other non-human beings felt, even though subjects new that it was merely a drug-induced experience. As Comings observes, the sense of realness that subjects experienced under the influence of DMT, "goes to the heart of the conflict between science and faith."[14] At the very least it should probably be seen as a cautionary flag when confronted by any sacred text purporting to be coming from someone claiming to be connected to God in some special way. On this, Comings notes that "it is simply necessary for the rational brain to understand that one of the characteristics of the spiritual

11. Ibid., 367.
12. Ibid., 368–374.
13. Barker, S. A., J. A. & Christian, S. T. N. N-di-methyl-tryptamine: an endogenous hallucinogen. *Int Rev Neurobiol.* 22: 83–110.
14. Comings, 383–384.

brain is to strongly believe in something and have faith in something, even when the rational brain says it is unreasonable or that it did not and could not have happened. The rational brain needs to recognize that the spiritual brain can sometimes confuse externally generated reality with internally generated experience."[15]

In addition to DMT, there are a number of other molecules that are similar to the neurotransmitters, and which produce similar results as does DMT. Among these are psilocybin, which investigators found long-term positive effects, resulting in improved attitudes and appreciation of life, deeper religious faith, and greater empathy for others.

The Natural Spiritual Brain

A number of studies indicate that observable changes occur in the temporal lobe associated with certain physical activities—namely transcendental meditation and charismatics. In reporting on this, Comings cites the work of several investigators, one of which was Michael Persinger, a professor of psychology at Laurentian University in Ontario, Canada. He references an EEG study as providing an answer to the question of a connection of these activities to the temporal lobe, noting a spike in the activities of this part of the brain, "during and only during protracted intermittent episodes of glossolalia." Also emerging from this same study were findings that delta-wave dominant electrical seizures emanating from the temporal lobe occur during peak experiences for routine transcendental meditation. The author of this study proposed that such occurrences were in his words, "consistent with his hypothesis that transient, focal, epileptic-like electrical discharges in the temporal lobe, without convulsions, may be associated with strong spiritual and religious experiences."[16]

Meanwhile, Persinger's research suggests that a person's "God belief" is composed of a combination of 1) God Experiences, and 2) God Concept, with the former being associated more with the emotional self, and the latter being associated with the cognitive or thinking self. It was Persinger's conclusion that experiencing the sacred is part of the natural construction of the temporal lobes, representing something on the order of its normal function, and not comparable to TLE or other extraordinary temporal lobe manifestations.

---

15. Ibid., 384.
16. Ibid., 394; see also Persinger, M.A. Striking EEG profiles from single episodes of glossolalia and transcendental meditation. *Percept Mot Skills*. 58: 127–233.

Regarding the experiential aspect, he notes the following:

> Usually the God Experience involves euphoria and positive emotions. The person reports a type of God-high that is characterized by a sense of profound meaningfulness, peacefulness, and cosmic serenity. Invariably the state is perfused with references to reduction of death anxiety. It is defined as the anticipated extinction of the self-concept or "the thinking entity." During the God Experience, the person suddenly feels that he or she will not die. Instead, he or she will live forever as a part or subset of the symbol of all space-time. If the symbol is a father image, then the person expects to become a child of the father. If the symbol is "imageless," the person expects to become part of a Universal Whole.[17]

It is Persinger's conclusion that such experiences are part of the natural construction of the temporal lobes, representing something on the order of its normal function, and not comparable to TLE or other temporal lobe manifestations.

There is a downside to this aspect of the natural brain. For Persinger it has to do with the fact that the human tendency is to assume experience equates with what is real and true. In his words, "We assume that if we perceive something, it must exist, or, if we have not experienced something, it does not exist."[18] Looked at in this way, a framework is established, by which it becomes possible to critique the holder convicted that they have cornered the market on truth. The powerful combination of both the *God Concept* and *God Experience* delivers too many individuals dogmatic certitude in the accuracy of belief and the validity of the experience. While some elevate and praise such convictions, history has made abundantly clear that they come with a dark side, for a lot of blood has been shed by people who believed they were advancing the work of God. In Comings' words, such individuals "may kill or sacrifice themselves for [some] benefit, or proselytizes others to believe that they have the only true belief system. They can represent the most rabid of the religious believers."[19]

Summary Thoughts

This chapter has merely skimmed the surface of this topic, with studies in this field ongoing. As such, it would be inappropriate to conclude that

---

17. Persinger, M. A. Neuropsychological Basis of God Beliefs. Prager; see also Comings, 390

18. Comings, 393.

19. Ibid., 392.

knowledge in this area of study has reached any sort of terminus point. For sure brain science is a complex field of study, with this chapter having provided readers with some sense of current scientific insights.

Many will likely recognize a tie between some of this recent research regarding TLE and the history of spiritual gifts within organized religion. David Comings boldly ventures a number of case studies, both biblical and extra-biblical, with the main idea being that the manifestations of many prominent oracles from religious history possessed spiritual gifts around which reasonable suspicions can be had of a very natural explanation. I have chosen not to go down this trail for the simple reason that it is somewhat outside the main thrust of this book, as well as the fact that some will view it as inflammatory given that there is no way to achieve clinical proof for the physical nature of spiritual gifts as ascribed to historic figures that no longer live among us physically. This reality will remain as a convenience to those who live with the faith that such manifestations are something more than natural pathology at work. The most that can be accomplished from a remote distance of time is to note surviving information that has a consistency with the symptomology represented in temporal lobe pathology. Nevertheless, the increasing clarity that is beginning to emerge regarding these physical manifestations represents a sober reality check for some of the more extreme sacred claims. In the face of an increased level of knowledge regarding natural brain processes and pathologies, it would seem that the assumptions that are frequently in play as to the origins of spiritual gifts among many religious leading lights probably deserves closer scrutiny.

For sure, a great many theologians recognize there to be a human component represented in all revealed texts. Temporal lobe phenomenon merely provides the logical vehicle by which some of these manifestations can materialize. So with this in mind let me suggest that there is an alternative way to respond to the findings outlined in this chapter that can maintain respect for some of these venerated oracles. First, it would seem appropriate to hold sympathies for anyone who may have operated honestly out of certain pathologies, as well as for the adherents who may have signed on to a program that was substantively not what they thought it to be. Second, to understand some of these manifestations to be the product of biological abnormalities can be a way of explaining very different perspectives of God that have emanated from these varied oracles who penned writings that have sometimes been deemed either explicitly or implicitly as words from God. The dissimilar portraits of divinity that sometimes come from these differing revealed sources can be more easily accommodated if it is recognized that "God" is to some extent always going to be a human projection when communicated through a human agent, and that such projections may in some cases be influenced by the results of an abnormal temporal lobe condition. Third, assuming the veracity of the prior points, it would seem a bit presumptuous to reach any conclusions about whom God uses in furtherance of his purposes, as there simply is no objective standard by which to judge such matters, other than 1) on the basis of whether the message is internally consistent, 2) its rough accuracy in the outcomes of prognostications about the future, and 3) a general concordance with empirical findings.

For now, it seems, it may be necessary to entertain the possibility that some of the successful religious visionaries may have been unwitting victims of TLE, a point of significance to followers who thought they had a more direct connection to the divine than other mortals. To the extent that this may be the case, it is probably worth facing the prospect that brain pathology has been responsible for creating a lot of diverse religious and human drama, as one belief system jockeys for influence or control over other communities of belief. It may well be that TLE factors not only in the formation of some of the great religions of the world, but also in the conflicts that inevitably have erupted between Judaism, Christianity, and Islam, not to mention those collaterally related whom have all suffered as a consequence. Surely there will be an ever-increasing clarity with respect to the role of TLE in the history of humanity and religion in the years ahead as the mystery of temporal lobe manifestations continues to unravel. For now it is simply important to be aware of this issue before choosing to die on a hill that may not be what it appears.

The Quest for a Theology Connected
to the Real World

CHAPTER 8

The Power of Story

Meaning is made and sustained in conversations. It lives in relationship: in marriages, families, communities and societies. It is told in narrative, invoked in prayer, enacted in ritual, encoded in sacred texts, celebrated on holy days and sung in songs of praise[1]

—JONATHAN SACKS

Chapter Summary

Humans live within the framework of mythologies and narratives and the assumptions upon which they are created. It is important to recognize that interpretation of reality in all of it permutations is never baseless apprehending. At a foundational level all thinking begins with a tacit knowing, from which a master narrative forms that may include a mix of superstition, reason, empiricism, and relationships from one to the other. It is interesting to note that all pre-reflections that mature into ideas of significance tend to become closed systems, as the starting assumptions naturally tend to close off other options. As purpose driven beings, humans have long been creators of narrative about their world, doing so in ways that can create a meaningful life. Such narratives maintain credibility as long as they remain respectful of other forms of knowing, only becoming an issue when making

1. Sacks, *The Great Partnership*, 2011.

assertions that step outside the boundaries of verifiable realities. It is here that religious convictions often run into trouble, for it is not uncommon to engage in interpretations that are totally oblivious to real world data. Scientists can also sometimes overstate what is knowable, yet generally live within the framework of methodological naturalism that includes a self-correcting process. If religious systems learned nothing more than the development of self-correction, it would help immensely in keeping them from veering off down dead-end rabbit trails. As such, modesty and humility can go a long way in keeping a sacred master narrative on a credible path.

In earlier chapters the central role that is played by the senses in achieving a grounded connection to the real world has been developed, and yet there may be some importance in understanding that such a source is not the only way in which people connect to the world. As powerful as the senses are in creating a basis for knowledge, they would be of little help in the absence of cognitive powers to organize sense data into some sensible framework of logic and reason. For many centuries, leading thinkers have debated as to which is more fundamental. Those who take the time to study the arguments for and against will likely discover that this is not a question easily resolved even though up to this point most of the focus has been on empiricism, and to a lesser extent on authority and rationalism. But just for the record, there are other sources of knowledge that some thinkers point to as well, and among these are subjectivism and pragmatism. There are scholars around, who sometimes argue that one or the other of these varied pathways represents the primary path to knowledge, and thus the most appropriate framework for a master narrative.

In sorting some of this out, I have found the thinking of the European scholar, Michael Polanyi, helpful. It may be recalled that he was referred to earlier. He was an individual well versed in the methods of knowing from the disciplines of both philosophy and science, and acknowledges in particular the value of both empiricism and rationalism. He gave significant consideration to the respective priority that should be accorded to sensory experience—something that he as a scientist would value very much—verse theory and logic. This was discussed at some depth in his book, *Personal Knowledge*.[2]

2. Polanyi, *Personal Knowledge*, 266. Here, Polanyi discusses the value of presuppositions in detail and why theoretical knowledge should be thought of as being more objective than knowledge acquired from the senses. It is a significant book and well worth review by those considering this issue. A nice summary can be found in Leroy Seat, Robert E Patterson ed., *Science Faith and Revelation*.

In the end, he argued that theory is more objective than sense data for the simple reason that the senses can sometimes fool an investigator, making the observation that rationalism offers assistance in figuring this out.

His poster-child in this argument was Ptolemy's geocentric understanding of cosmology. Readers may recall that this model proposed the earth to be at the center of the rest of cosmology, with all the heavenly bodies revolving around planet earth. In fact, when looking up in the sky, this is the feedback received. Though the folly of this view is now universally recognized, geocentricity was based on solid observations that had a seducing effect. In short, a geocentric view of the universe was science—based on observations. When Copernicus came along and proposed the heliocentric model with the sun at the center of cosmology, this alternate explanation was being proposed on the basis of rationality and simplicity instead of solid data to fully support the proposal. Yet his idea did resolve certain anomalies of the Ptolemaic system involving the mysterious epicycles where the planets sometimes appeared to be moving in retrograde motion. It wasn't until Galileo discovered supporting evidence for the Copernican hypothesis—the timeline of which came sometime later—that this idea was validated.

It was from this case study that Michael Polanyi argues the priority of reason—it seemingly being more objective than sense data. He supports this by pointing out that it was theory that led the way to resolving this issue, given that the senses were suggesting a different conclusion.[3] He has not been the only one thinking along these lines, but he was certainly among its sophisticated expositors.

Polanyi differentiated between what he calls *explicit knowledge*, and that which he termed *tacit knowledge*. His use of the term *explicit knowledge* is somewhat analogous to what was noted earlier about a type of knowledge that can be characterized as having some categorical qualities about it that elevate it well above that of mere opinion. Alternatively, he identified tacit knowledge to be non-verifiable and a-critical. It was Polanyi's contention that there is a tacit component present when deciding what constitutes evidence for a hypothesis. He notes, for example, that *evidence* does not come labeled as such in nature. Evidence derives its label only to the extent that it is defined that way by the observer, and others have made much the same observation as has been noted in the earlier discussion. The validity of this reality is often witnessed where there are conflicting sets of presuppositions because as he notes, "the two sides do not accept the same 'facts' as facts,

---

3. Ibid., 336–354.

and still less the same 'evidence' as evidence.[4] Well, this is basically the point many social constructivists have tried to make.

He went on to note that theory itself has a subjective element in its starting point due to the impossibility of ever getting outside of self or culture to make objective judgments about anything. It was his view that knowledge must be guided by "antecedent belief." What he meant by this was that the start of the thinking process begins with a set of givens (belief). While belief has long been thought to be inferior in its validity to empiricism, Polanyi held that belief is the source of all knowledge.[5]

In support of this he made much of what he termed "local rootedness," and by this he was referring to the presuppositions that a person commits to prior to thinking. He indicated that behind all knowledge there is the pre-logical and a-critical commitment to certain implicit beliefs, observing that there is no way to test or verify a basic set of presuppositions to which a person commits. As he puts it, "we live in the garment of our own skin." These presuppositions are transmitted from one generation to another, and a child grows up committing to them. Thus, tradition is indispensable: "a society which wants to preserve a fund of personal knowledge must submit to tradition."[6] This all provides the possibility of religious belief playing a role in the systematic decoding of the laws that run the universe.

Charles Taylor makes much the same point in his tome *A Secular Age* in which he notes that all reasoning is shaped significantly by a pre-reflective grasp of the world, noting that: "An interpretation is never presuppositionless apprehending of something presented to us."[7] In addressing Charles Taylor's thesis in his subsequent dissertation, Zane Yi attempting to clarify this point by making an insightful summation, noting that, in his words—" . . . there is no view from nowhere and there is no view from everywhere. Humans are affected by the societies they live in, the states that govern them, and the families that rear them. These factors have 'a pre-determinate influence' on any experience an individual may have. Humans are finite and historical beings."[8] He then goes on to offer up an analogy of a piano concerto, it being composed of a multitude of notes arranged in a

---

4. Ibid., as to Polanyi, 53–64, 167, 264–268.

5. Ibid, 266. Polanyi was early to the present day debates related to social constructivism which I have not gone into in this book, but which covers a rather wide spectrum of thought with the extreme end of it tending towards nihilistic and post-modernist thinking, often denying the existence of objective reality. Regarding issues of social constructivism, see generally James Robert Brown, *Who Rules in Science*.

6. Ibid.

7. Heidegger, *Being and Time*, pp. 191–192.

8. Zane G., *The Possibility of God*, a dissertation, Fordham University, 69.

given configuration. If one single note of that concerto is played it can be experienced, but it has little meaning outside of some context—that being the experience of the notes than came before, as well as anticipation of the notes that come after. His point being that humans create meaning from the memory of that which came before (human history and traditions) and by anticipation of what can be expect in the future.[9]

Master Narratives

At this juncture it may be useful to consider the conversation narrative that plays in the head of every individual, ever-evaluating and contextualizing all that happens in the world of the personal, social, political, and theological. Such narrative provides a basis for analysis, but can also contribute texture. It is in that way that the behavior and beliefs of others gets evaluated, and narrative kicks in to fill whatever level of ignorance may be in play so as to explain what has transpired. This means that sometimes these knowledge gaps will require guesses, or assumptions. Occasionally judgments, based on only bits of information, may be made that are relatively on target either by luck or by mature perception, while at other times they may be strategically off base.

This ongoing narrative provides context and meaning to life, and is ultimately guided by a worldview. As beings that live within the framework of story, understanding its source is important—including history and context, politics and societies, religion and related traditions. The strengths and weaknesses of each of these sources becomes an important part of the analysis. While all of these sources contribute to story, perhaps the most important and influential would be the broad narratives that develop from the varied religious traditions. As purpose driven beings, humans have long been creators of narrative about their world, doing so in ways that can create a meaningful life. Even fairly primitive cultures have stories that attempt to explain reality and provide meaning, purpose and structure to life. They afford a framework from which events that materialize are given some context of meaning, and then become a part of the symphony of notes that give meaning to the specific notes of life.

Religion, which falls under the rubric of mythos, offers up a master narrative built on certain traditions passed down through the generations, and in the case of theism, is based on the insights of individuals who either claimed to have a divine connection, or a tradition that has imputed such through contemporaries or others who came afterward. The problem this

---

9. I was part of a small group in which Yi made this analogy in describing the major outlines of his dissertation, January 11, 2014.

represents for those who seek to connect such insights to sensate and rational endeavors has to do with validation. There are, in fact, a very large number of so-called sacred texts and traditions, and unfortunately in a number of important respects they are not in accord with each other on many points—something that has already been considered to some extent.

This reality creates a major obstacle for attempts at objective inquiry, for by what criteria should they be individually judged?

It was noted, for example, in an earlier chapter that the Jewish Scriptures composed of the Old Testament documents present a different picture of God in certain important particulars than those represented in the primary Christian documents, that being the New Testament. This issue led Marcion in the second century to reject the Hebrew Scriptures, given the Hebrew understanding of God was so starkly different from the understanding that emerged from the Gospels. As readers know, Christianity ultimately adopted both Testaments, even though controversy remains to this day. While some are willing to jump through hoops defending the equal validity of all views of God represented in Scripture, many scholars are mindful that this issue is not so simply resolved. If this reality weren't bad enough, by expanding the focus to more universal considerations that might include many other so-called sacred writings the issue becomes even more acute with many other writings considered by various adherents to be inspired or sacred, each representing certain incompatibilities with the others.

The task of approaching varies sacred writings objectively is formidable, for the question again is, "by what criteria should assessment be made as to the validity of one over the other?" Where should the investigator begin, and by what measure should some of these problems be untangled so that they may have the capacity to enhance understanding of the larger reality, and doing so in a way that can separate the helpful revelation from that which fails to make such a contribution? For sure, unless there is some basis of grounding claims of the sacred with the empirical world the task will be difficult to impossible.

All of this illustrates the point Polanyi and Taylor have made, namely that some of the most important beliefs an individual may hold grow out of the pre-logical, they being adopted in an a-critical nonsystematic manner. It comes by way of an inheritance—from parents and culture that tracks back to the fog of antiquity. Thus, a Muslim likely had Muslim parents; a Christian—Christian parents. To acknowledge this fact does not mean there aren't exceptions, but in general children embrace the worldview they inherited by birth. Furthermore, most do not give a great deal of objective thought as to why they have adopted one particular identity verses another. As an inheritance, the common practice is to simply embrace a spiritual heritage and then defend it against outside assault. As such the power of an inherited master narrative comes in its ability to transmit values by offering the adherent some context for understanding the world—it representing a fundamental part of human identity.

This process characterizes how the empirical world is differentiated from the world of mythos—the former being connected to the senses, the latter being connected largely through tacit forms of knowing, and private evidence—prophets, sacred texts, and traditions. Such sources are frequently an inheritance of birth. Scripture plays an important role in this for the Judeo/Christian tradition—the Qur'an for Muslims. In all of this, mythos goes beyond where science is capable of going by addressing questions of purpose, meaning and values. It explains the inexplicable and creates an important context for worldview.

While some may make the mistake of understanding mythos as representing a class of ideas that are untrue or in some way a part of non-reality, such is not the necessary conclusion. In actual fact, such ideas and beliefs should not be mistaken in this way for the term itself makes no judgments about correlation of any particular master narrative to reality. The three inescapable conclusions would be the following:

1. Mythos is very powerful in providing purpose and meaning;
2. That there is a very large assortment of variations contending for adoption as narrative;
3. That obviously since many of them conflict with each other they cannot all have a concordance with reality. So by logic it becomes clear that mythos cannot be relied upon for having a necessary correspondence to reality, though many assume the mythos version they have adopted has an exact concordance, and then proceed to call it *truth*.

Tension can arise when competing mythologies vie for space in the marketplace of ideas, or intersects empirical data that seems to be saying something different. On such occasions this becomes the junction in the road—for mythos it becomes an opportunity to grow up and be open to the outside world, or alternatively to risk retreating into irrelevance; for the science community it becomes an opportunity to recheck the data to make sure it is being interpreted correctly.

The process of putting mythos on a more objective footing can be assisted in the process by turning to the insights of the empirical world, where certain myths intersect with data that can offer grounding for aspects of "story" that can be validated or falsified by data. As an example, it can be helpful to bring scientific insight into any reading of the early chapters of

Genesis. Obviously, science by proceeding on the basis of methodological naturalism cannot validate the idea of God as creator, but certain other elements can be analyzed. Current scientific thinking validates the universe as having a beginning; that the appearance of life had a progression to it; and that the appearance of humans were rather late in the progression. Thus, if Genesis 1-2 is not over-literalize, much of the overarching ideas it presents are very much in sync with science—something that is rather amazing considering the fact that it was written some 2500-3000 years ago. On the question of theism itself, while not within the domain of science specifically, there are certain aspects of inquiry that tend to bolster the idea, and some of these ideas will be considered later.

Mythos can also be advanced when it provides the widest deference to its sacred text by taking critique to a strategic level rather than wading into the tall weeds a of reductionistic belief system. A strategic approach attempts to capture the overall essence rather than getting bogged down into some of the problematic details. This will be discussed in some specify in chapter 10.

Evaluating Master Narratives

No master narrative is above critique, yet the critique process can be difficult. Not only does it come with a great deal of emotional attachment by the holder, but also as already noted they are constructed on a foundation of assumptions—those tacit elements of knowing to which Polanyi and Taylor speak. This reality complicates critique. Nevertheless, there are at least three factors that come to mind that could be helpful to the critiquing process. First, it would seem important to reflect on the internal logic of the narrative itself, and to the extent that it represents an issue, can become a point of legitimate critique. Second, if a narrative is based on one or more conflicting assumptions, such would be a basis of critique for the simple reason that logic militates in favor of resolving inconsistency. Finally, if certain assumptions are in conflict with data, the narrative holder might be inclined to act dismissively towards the interfering elements by essentially engaging in a process of cherry picking—latching on to supporting data and ignoring the damning data. Yet, such conflicts should serve up notice that this may be an area needing further study—not dogmatism. At the very least it should represent to the narrative holder that the matter requires humility of spirit.

Earlier discussion noted the fact that some assumptions about the beginning of this world are based on certain interpretations of Scripture, even while some of these understandings are in direct conflict with a sizeable body of scientific evidence. This reality should put people on notice that this represents an occasion for caution. There is always a chance that the data has been misinterpreted, though this possibility diminishes as multiple streams of independent data essentially lead to the same conclusion—that being the appearance of a world that has a beginning that extends into deep time. In whatever way such issues are resolved it is important to recognize the dangers of expressing certainty in the face of data that is interpreted in a contrary direction—particularly in light of the problems that have been discussed regarding all sacred texts.

Central to all of these issues is the importance of a master narrative to self-identity, given that it controls how each person sees and interacts with the world. Perhaps the best illustration of this would be the assumptions that are made about ultimate reality, an example, of which involves the question of God being the creator of the universal order. If this question is answered in the affirmative, it will naturally have a profound effect on one's worldview and the thinking that flows from it gets ordered accordingly.

Given the significance of narrative to personal identity it is understandable that it often comes with baggage, including occasional resistance to data that may represent a challenge. Earlier, mention was made of Charles Taylor who initiates the term *closed world structures (CWS)*, and by this terminology he was referencing, in a very narrow way, pre-reflective systems of thought that exclude the transcendental (God).[10] Interestingly he does not attempt to argue for a transcendent reality, but merely offers critique on those systems that would exclude the possibility. This got me to thinking that while the terminology CWS seems descriptively useful, perhaps it should be expanded, for surely all master narratives are closed world structures to a greater or lesser extent—not just those that exclude the transcendental. Whatever belief system is embraced, whatever starting assumptions are in play—other options are necessarily being precluded. In this sense, all humanity tends to live in a closed world.

A contemporary example of closed world thinking that most people will be familiar with comes with the political chattering class in the United States. Major segments of American media have abandoned any pretense of objectivity by creating narratives that fit a strictly partisan mold—and this happens both left and right. Many of these dialogues eagerly tailor the facts in a cherry picked sort of way, or they conveniently omit facts by building subjectively partisan arguments to advance a perspective. Largely gone are the days when commentators put all the significant facts on the table, allowing viewers to figure out the reality.

---

10. See Yi, *The Possibility of God*, 9—10.

Those who embrace one closed world or the other seldom recognize the procedural malfeasance of those partisans they identify with; seldom is there any conscious recognition that one lives within a closed world. Consequently, there often exist parallel universes of partisan dogma, complete with a set of facts that are largely ignored by the opposing parallel closed world.

This reality also exists in the realm of the sacred. Those who choose the Judeo-Christian master narrative close out other potential options—whether naturalistic or theistic in substance. Furthermore, as consideration moves to denominationalism, it becomes apparent that narratives can form within a more global narrative. This, of course, is somewhat common to all master narratives, and unless there is some respect for objectivity, with some commitment to be open and fair-minded, a lot of evidence will be overlooked for the simple reason that it does not fit nicely into the operative narrative.

In the realm of revelation, a more open and balanced approach must recognize the significant role that both empiricism and rationalism each bring to the human quest for wisdom about the universe. Obviously humans will always be forced to live with a certain amount of abstraction as to exactly where to place the balance, though it seems that Polanyi has made a rather compelling argument in favor of reason, where the tacit avails a role for sacred constructs. The dance of ambiguities involved in all of this will be quite unsatisfactory for a lot of individuals as fundamentalism so clearly illustrates—where the demand for concrete certitudes dominate regarding ideas that are not inclined to be so accommodating. Yet, to understand that ambiguity is all part of the messiness of what it means to be human must be a point to be reckoned with—not because it is desired, but because there is no other choice for thinking sensate beings.

The existential reality is that humans have no choice but to live with the hiddenness of God, with the avenues for making that connection afforded through sacred writings that are often ambiguous, conflicting, and of problematic provenance. In spite of these realities the human spirit moves forward through story. In the life of every human it plays a critical role in making existence meaningful. It is, in fact, mythos that provides a sense of purpose and meaning. In this respect it creates for the holder a way of viewing the world, providing context for how to understand the world.

Summary

For those unfamiliar with the process-of-knowing arena, perhaps something more should be said about why it is of such importance. Essentially all rational thought must have a starting place—a set of givens or things that are assumed prior to any serious systematic thought, but which may not be testable in reaching conclusions. Some assumptions are conscious, and some may be unconscious. Examples of what might be include on a short list of conscious assumptions would be the following:

1. That there is a real reality out there and it is accessible through the senses;
2. That there is a level of know-ability about reality—otherwise the scientific enterprise would be pointless;
3. There is order to be found in the world through observations of cause and effect, and humans are capable of discovering it through their rational faculties. This is generally done by working off of a hunch and then forming a hypothesis that is then tested;
4. The order found in nature is reducible to a set of laws and principles. These laws and principles create a predictable world because they are repeatable and can be tested with reproducing results.

There are other presuppositions that could be outlined, but these would at least be on most people's lists.[11] Absent from the above points is anything that would suggest a creator of the universal order, something that is generally assumed by those from the faith community. It seems entirely appropriate that it would be include as long as it appears in the proper forum. Certainly such assumption is inappropriate for any science-based methodology that proceeds on naturalistic terms requiring a sensate foundation. This would extend to the educational science curriculum—which some have attempted to infuse with theology under the guise of Intelligent Design. The reason for this is that, as has already been noted, science proceeds with a starting point of reliance on the senses, and essentially what that means is that it does not presuppose theological ideas because such is not available to the senses.

Most Christian scientists I am acquainted with are completely comfortable with this naturalistic approach, for how can science make progress if that which is not understood is automatically attributed to an unseen deity. Thus, science, in this sense is constrained and agnostic on such question. On the other hand, theology is not so constrained. Consequently, the tacit component of which Polanyi speaks may offer a way forward for those who hold to sacred presuppositions, yet at the same time recognize that the contribution science has to offer is not free to take this course. It is specifically

---

11. Even though these are fundamental it must be acknowledged here that some philosophers would disagree with some of these presuppositions. Nevertheless, in the main, most scientists would consider these presuppositions to be on solid ground.

on this point that I find the thinking of Polanyi and Taylor insightful, reinforcing the major role of the tacit component given that it provides a pre-reflective grasp of the world—in the words of Zane Yi— " . . . there is no view from nowhere and there is no view from everywhere."

This chapter has attempted to convey how all humans live within the framework of narrative, with the assumptions upon which it is created shaping worldview. Revelation, reason and sense-based data would seem to be the most common components for worldview thinking. As powerful as the senses are in creating a basis for knowledge, they would be of little help in the absence of cognitive powers to organize sense data into some sensible framework of logic and reason—all beginning with the tacit component of starting assumptions. It is also the tacit component that can allow space for revealed authority that moves beyond descriptive narrative to address questions of purpose and meaning.

In the next chapter, consideration will be given to some of the human attempts to develop the most credible way in which to address sacred—matters of ultimate concern. A variety of very creative efforts have been devised for creating sacred narrative in a more credible manner. Some have been very helpful—others less so. In any event, a strategic sense of how these ideas are developed can provide assistance to those who seek to create a more credible master narrative.

CHAPTER 9

Paradigms of Meaning[1]

Chapter Summary

This is a review chapter that looks at some of the common philosophical/theological systems that have attempted to provide validation and support for a transcendent reality. There is too much material to go into any detail, and probably the best that can be done here is provide a listing of the systems of thought that have been mentioned. They are as follows: a) arguments from religious fidelity and experience; b) theological rationalism; c) theology of causation; d) pragmatism; e) natural theology; and f) agnostic arguments. In the end, none of these approaches can move knowledge of God to an explicit level, yet in certain ways many of these ideas can be helpful for those who desire to live life with a master narrative that assumes God, adding meaning and purpose to life. The goal of all these approaches outlined below is designed to elevate the idea of God above the plane of mere opinion, and while none are completely successful, they can provide synergistic support in that direction. The *master narrative* that guides each person through life plays an important role at this juncture, for it matters significantly how it is framed as to whether it will be capable of creating a purposeful existence. Perhaps the most significant aspect of this narrative is that control remains with the individual. Ultimately, while it is not possible to escape finite human capacity, there are many avenues by which a master

---

1. In the title of this chapter the word "paradigm" is used. Sometimes in the sciences, it has a fairly specific meaning for a theoretical framework formulated from theories, laws, and supported by generalizations from experiments performed. More broadly, it is simply a theory or a group of ideas about how something should be done, made, or thought about. It is in this latter context that it is being used here.

narrative that includes God can be put on an elevated footing. It comprises a number of features that attempts to exploit the powers of sense and reason, and in some cases to move beyond these standard tools.

⸺

In several of the preceding chapters the discussion has focused on the important role that the senses play in bringing some level of knowledge. Certainly the gift of sense, when combined with reason, provides for a level of intimacy with the universe that would not otherwise be possible. Yet as magnificently effective as this process has been in elevating knowledge of the cosmos on so many levels it is difficult not to miss its greatest weakness, namely its inability to speak to the core issues of purpose and meaning mentioned in the last chapter in the discussion of narratives. The potential role it can play comes with recognition that methodological naturalism has been unable to resolve the existential puzzle, and few hold out any hope that it will ever by successful on this score. The natural flow of all human inquiry regarding the natural order ultimately flows to questions of meaning. In the introduction we quoted Philosopher Alan Watts, who spoke eloquently to the dilemma of the modern secular world. His statement is sufficiently profound that it bears a second look:

> By all outward appearances our life is a spark of light between one eternal darkness and another. Nor is the interval between these two nights an unclouded day, for the more we are able to feel pleasure, the more we are vulnerable to pain—and, whether in background or foreground, the pain is always with us. We have been accustomed to make this existence worthwhile by the belief that there is more than the outward appearance—that we live for a future beyond this life here. For the outward appearance does not seem to make sense. If living is to end in pain, incompleteness, and nothingness, it seems a cruel and futile experience for beings who are born to reason, hope, create, and love. Man, as a being of sense, wants his life to make sense, and he has found it hard to believe that it does so unless there is more than what he senses—unless there is an eternal order and an eternal life behind the uncertain and momentary experience of life-and-death."[2]

Outlined here is the apparent reality that evidences itself by using the standard tools of investigation. It offers up little hope to sentient beings who long for purpose, and seek something beyond a transitory existence.

2. Watt, *The Wisdom of Insecurity*, 13.

This situation creates a tenuousness that philosophers have attempted to bridge through the use of reason, endeavoring to establish a rational basis for transcendent belief. Because these efforts can be rather significant to a life of meaning, there would seem to be some value in considering some of the significant scholarship projects that have developed over the past many hundreds of years. It perhaps can be characterized as the formal and systematic human questing after God. The topical areas that will be considered below are neither exhaustive nor detailed. But in an effort to explore the range of what is knowable through some combination of reason and exposure to the details of the world, this discussion is important. Furthermore, it may contain philosophical developments that some readers may not be entirely familiar with. Thus, some of the more significant and noteworthy ideas that have emerged will be considered as sort of a sampling of the range of thought that exists. As will become apparent, some of these ideas overlap.

Arguments from Religious Fidelity and Experience

In the community of faith in which I was reared, there has been relatively little official thought given to some of the problems that are conveyed by incorporating the words *God* and *knowledge* into the same sentence.[3] In many respects this community of faith would seem to be a microcosm of the larger community of faith that compose the spectrum of religious belief systems that span the globe where belief is basically a function of revelation combined with faith. Simply put, a very large number of individuals within the "faith community" do not critique the details of belief, which puts a superficial veneer on whatever meaning is thereby derived.

It is a short step from cognitive belief or its experiential aspects to the projection of God's existence. In its strong form, this argument asserts that it is only possible to experience that which exists, ignoring of course the possibility that a projection of mind can seem very real. The somewhat more nuanced version of this argument asserts merely that religious experience itself constitutes evidence for God's existence, though it is probably a good idea to recall the role of the temporal lobe in all of this.[4]

---

3. I would exclude from this assertion a number of very thoughtful scholars that it has been my privilege to know. This comment is directed more at the ecclesiastical hierarchy, whom have an institutional disincentive to verbalize doubt, preferring to reassure the flock with declarations of certitude.

4. See http://www.philosophyofreligion.info/theistic-proofs/the-argument-from-religious-experience/.

Perhaps the most significant shortcoming with this approach is the dilemma of conflicting experiences. In considering the full spectrum of religious claims made in the market place of religious ideas, it becomes quite apparent that many conflicting claims are based on revelation, faith and experience. So, unless some objective criteria can be established that could address these varied understandings, it becomes difficult to assign a lot of credibility to this method in spite of the fact that many honorable people approach faith commitments in this manner.

Theological Rationalism

Ontological arguments seem to generally be a little more sophisticated than those based on religious fidelity or experience. This variety of argument falls under the umbrella of rationalism, basing arguments of God on logic alone, without any appeal to sensate evidence. In its classic form it attempted to prove God's existence through abstract reasoning, though many disciples of this approach acknowledge the possible infinite regress of such proofs and conclude that there must be truths or beliefs that can just be accepted and reasoned from; sort of what can be thought of as starting presuppositions.

St. Anselm, Archbishop of Canterbury (1033–1109) was the originator of this argument, with it being summed up as follows:

1. God is a being that which no greater can be imagined (that is, the greatest possible being that can be imagined).
2. God exists as an idea in the mind.
3. A being that exists in reality is greater than a being that exists only as an idea in the mind.
4. If God exists only as an idea in the mind, then we can imagine something that is greater than God (that is, a greatest possible being that does exist).
5. But since we cannot imagine something that is greater than God, it is therefore concluded that God exists."[5]

In more recent times the ontological argument has been enhanced by use of modal logic, distinguishing between necessary and contingent realities. One of the early exponents of modal thinking as applied to God was Kurt Gödel, best known for his incompleteness theorem. He developed a logical proof of God using an ontological modality, distinguishing between necessary

---

5. See http://www.iep.utm.edu/ont-arg/.

and contingent truths. From a set of axioms Gödel argued that in *some* possible world there exists God. Gödel also discusses and defines *essences*:

> If x is an object in some world, then the property P is said to be an essence of x if $P(x)$ is true in that world and if P entails all other properties that x has in that world. We also say that x *necessarily exists* if for every essence P the following is true: in every possible world, there is an element y with $P(y)$.[6]

He then argues:

> ... that since necessary existence is positive, it must follow from Godlikeness. Moreover, Godlikeness is an essence of God, since it entails all positive properties, and any non-positive property is the negation of some positive property, so God cannot have any non-positive properties. Since any Godlike object is necessarily existent, it follows that any Godlike object in one world is a Godlike object in all worlds, by the definition of necessary existence. Given the existence of a Godlike object in one world, proven above, we may conclude that there is a Godlike object in every possible world, as required."[7]

As astute readers may note, this approach parallels the original ontological argument first proposed by St. Anselm by extending the mathematical probabilities of the statements made. Certainly as modalism applies to theology, one of the biggest objections is that it permits the rational acceptability of almost any belief; and some claim that it smacks of fideism, the view that "belief in God should be held in the absence of, or even in opposition to, reason."[8]

Alvin Plantinga is among those who favor a modified version of the ontological argument. He employs what he refers to as "reformed epistemology" which views religious belief as foundational, and as such not needing of any external justification or evidence; it just needs to be rational, grounded in perceptual religious experience. Plantinga claims that humans have "a disposition or set of dispositions to form theistic beliefs in various circumstances, in response to the sorts of conditions or stimuli that trigger the working of this sense of divinity."[9] He refers to this as *sensus divinitatis*,

---

6. http://en.wikipedia.org/wiki/G%C3%B6del's_ontological_proof#Outline_of_G.C3.B6del.27s_proof.

7. Ibid.

8. See http://www.iep.utm.edu/relig-ep/.

9. Plantinga, *Warranted Christian Belief,* 173.

a concept in which the divine sense is direct, and not the result of empirical data, inference or other methods of rationality.

It appears that he is treating this approach as a beginning assumption or presupposition, from which all other thinking flows. As noted early, beginning assumptions are notoriously difficult as a category because they often represent the pre-logical, but also partly because everyone has them as a starting place of thought. The inference being suggested by Plantinga is that religious belief is foundational and not open to critique because it is the centerpiece of a master narrative that is derived from a personal sense of the reality of things.

When comparing the ontological approach to that of belief and experience based on faith and revelation, there is arguably an elevated level of sophistication involved; yet it is clear that they share some commonalities, namely that evidence for belief matters very little. As a result, this approach falls flat for a lot of people, including many philosophers and theologians who believe that evidence does matter in such discussions.

Theology of Causation

The cosmological argument holds that the existence of the universe is itself strong evidence for the notion that there must be a God who created it all. A general summation of the approach goes something like this:

1. The empirical observation is made that things exist.
2. Things that exist have a cause of existence, since something cannot bring itself into existence.
3. The chain of cause and effect cannot be an infinite regress, otherwise there is ultimately no initial cause.
4. Therefore, there must be an uncaused cause of all things.
5. The uncaused cause must be God.[10]

The strength of this argument is connected to both its simplicity and comprehensibly. It gives voice to the logic that objects from human experience do not bring themselves into existence and must, therefore, have causes.[11]

The Cosmological Argument comes in several forms, but mention will be made of just a couple of the more prominent of these forms. The first of

10. See generally Matt Slick http://carm.org/cosmological-argument; see also David Veksler, The One Minute Case against the Cosmological Argument, http://oneminute.rationalmind.net/cosmological-argument/.
11. Ibid.

these is the temporal argument—from first cause—distinguishing between the universe and God—with the universe operating within time and God being outside of time. The second approach emphasizes Modality—or contingency—as distinguished from that which is necessary (discussed above). It does so by making a distinction between the universe and God, with the universe being contingent, and God necessary.

There are several objections raised to this genre of argument. One objection holds that if all things need a cause to exist, then seemingly God Himself must also need a cause to exist, implying that there must be an infinite number of causes. One way this objection is countered is by distinguishing between things that operate within time verses those that don't (God)—or between things that are contingent, and things that are necessary.

Pragmatism

Blaise Pascal (1623-1662) was a seventeenth century mathematician who made contributions to both philosophy and theology. He arrived on the scene during the formative period of probability theory, coincident with the ascendancy of skepticism towards theism. He recognized that probability theory could make a contribution to issues that arise in theology, with him perhaps being best remembered for his argument that came to be termed as *Pascal's Wager*.

The essential idea pertains to the query that can be made as to God's existence. On this matter he noted, either God exists, or he doesn't. Given these two options an individual must wager one way or the other, and his argument was that the benefits of a wager in favor of God far outweighing any benefit of the alternative. He noted that if God does not exist, then it doesn't matter how we wager, for death will bring neither reward nor punishment. However, if God does exist, the only chance of winning eternal happiness is to believe, and the risk otherwise would be to miss out on this opportunity.[12]

He went on to refute agnosticism, arguing that it is not sufficient to sit on the fence of not knowing. While the agnostic may argue that "the right thing is not to wager at all," Pascal's reply was, "But you must wager. There is no choice. You are already committed." What he apparently meant by this is that we are not outside observers of life, but participants.[13]

Over the centuries since Pascal passed from the scene, his argument has been criticized on a number of grounds. Among the objections are the following: a) It runs into problems in the case of polytheism; b) it only

12. See generally http://peterkreeft.com/topics/pascals-wager.htm.
13. Ibid.

provides for a generic form of theism, and does little to improve on a more refined decision making; and c) it has also been critiqued for its calculating approach, with focus on self-interest.[14]

Some who have favored this framing conceptually have proposed a two-step combination that includes both pragmatic and epistemic features. It would seem that the pragmatic element is more directly related to Pascal's Wager. Once a person wagers in favor of God, then there are a number of arguments that can be pulled out that tend to make the wager more credible.

There are critics of the Pascal approach, namely that it is too calculating. It is not difficult to have certain sympathies for this critique, yet it can be helpful to remember that Pascal was really addressing some of the realities current in his world, namely a world that had increasingly turned its back on theism generally and Christianity specifically. Some of this was no doubt a reaction of disgust over the many excesses of religion recorded in the preceding centuries. To describe that period as the "dark ages" seems appropriate, given the savagery of the Crusades—all in the name of God. Excess often is followed by reaction, and in this case, many simply chose to throw the entire package of theism overboard. From this context, Pascal can be seen as reminding folks that it might be wise to make pragmatics a part of their decision-making matrix.[15]

Natural Theology

Natural theology comes in a variety of forms, but essentially attempts to approach the question of God from some combination of experience and reason, over against revealed sources. One major proponent of this approach is Richard Swinburne, Emeritus Professor of Philosophy at the University of Oxford, with formal academic acquaintance with the sciences. He attempts to show that theism is more probable than not, based on some relevant pieces of evidence, including: 1) the existence of the universe; 2) order in the universe; 3) the occurrence of conscious beings—these among others.

On the back cover of his book titled, *Is There a God?* he offers up the following observation:

> Surely the most natural state of affairs is simply nothing: no universe, no God, nothing. But there is something—and so many things. Maybe chance could have thrown up the odd electron. But so many particles!" What could account for the existence

14. See generally http://www.iep.utm.edu/pasc-wag/.
15 Ibid.

of such a rich and plenitudinous universe? And what could account for its many surprising features—notably its spatial and temporal order, its fine-tuned fostering of life and consciousness, its suitability as a theater for human action? There is a complexity, particularity, and finitude about the universe that cries out for explanation.

For Swinburne, the simplest hypothesis that explains the existence of our universe is that God is behind it. Readers may recall from chapter 3 in the discussion of the scientific method that the law of parsimony rules when it comes to explanations, and for Swinburne, the simplest hypothesis for explaining it all is to posit God. He notes that if science did not rely upon the simplest hypothesis, that they would never be able to get beyond the data. "To abandon the principle of simplicity would be to abandon all reasoning about the external world." Again he notes that "on a priori grounds, a simple universe is more likely than a complicated one. And the simplest universe of all is the one that contains nothing—no objects, no properties, no relations. So, prior to the evidence, that is the hypothesis with the greatest probability, is the hypothesis that says there is *nothing* rather than *something*."[16]

Unlike those philosophers who have attempted to prove God's existence by abstract deductive logic, Swinburne has used scientific reasoning, to show that the God hypothesis is more probable than not, doing so with the help of mathematical logic by way of Bayes' theorem—a statistical method that assists to demonstrate the probability of God's existence, verses nonexistence.

So that readers will have a sense of what a Bayesian approach is all about, the broad outlines has to do with it being a conditional probability, involving the calculation of a probability within a probability. This statistical technique is used in a wide variety of circumstances one of which has been in the area of paternity testing—a test that begins with consideration of the prior probability that someone who has been named the father of a child is in fact the father. In such cases, the alleged father is assigned ahead of time a 50 percent chance of being the father, thereby ignoring "he said," "she said" situations. The secondary part of the test considers the probability of shared alleles between the alleged father and child in question, with the overall approach being considered highly reliable. So it is in the use of this approach that Swinburne argues that introducing additional levels of evidence can increase the odds for the existence of God. This includes not just the existence of the universe, but also its lawfulness, the patterns of its

---

16. See Jim Holt, *Why Does the World Exist?*, 96. Quote taken from author interview with Swinburne. see also http://www.nytimes.com/2002/05/11/arts/so-god-s-really-in-the-details.html?pagewanted=all&src=pm

history, the nature of Jesus' behavior during his lifetime and the quality of witness testimony after his ascension. His conclusion is that it is more likely than not that there is a God.[17]

The assignment of prior probabilities is the most controversial part of the Bayesian method. While some are comfortable with the use of prior probabilities by arguing that these are logical matters that have their genesis in universal or neutral criteria, others argue that essentially these are subjective opinions. Swinburne rejects the idea that this method is subjective, arguing instead that it is important to select a theory prior to the consideration of evidence—sort of on the order of a hypothesis to be tested—which amounts to estimating its prior probability as higher than other possible hypothesis.[18] The elements that he would incorporate into such analysis would be the following: fitness with background knowledge, scope, and simplicity.

The initial steps in Swinburne's approach is not completely dissimilar from that of Michael Polanyi, referenced above, who made much of explicit knowing, verses what he termed as tacit knowing—those elements of thought that are assumed ahead of time that incorporate the pre-logical and a-critical.[19] So, for Swinburne, it is at this level that he considers what he believes to be the simplest of all hypotheses, that being the assumption for the existence of God. From there, one would go to the next level which would include consideration of points that would either contribute to, or negate the hypothesis. Nevertheless, it should be noted that some remain unconvinced that a "simplicity" argument is adequate outside of some definition of what "simplicity" entails, but also some basis for asserting that theism is necessarily the simplest concept. At the end of the day, it is arguable that Swinburne does demonstrate that there is a logical basis for thinking about God in rational terms.

Beyond the Swinburne approach, natural philosophy can also take on a more teleological framing with one prominent example pertaining to means and ends considerations involving that of the design and order found within the universe. Saint Thomas Aquinas argued from this angle, as did William Paley who came well after Aquinas. Paley, largely known for his 'watchmaker' argument, has articulated an idea that has resonated down through the ages of time and even today is considered by some to have put his finger on something that is important. His argument went like this:

> In crossing a heath, suppose I pitched my foot against a stone, and were asked how the Stone came to be there: I might possibly

---

17. Ibid, see generally, 92–104 where Jim Holts discusses his visit with Swinburne.
18. See generally, http://sammelpunkt.philo.at:8080/1679/1/portugal.pdf.
19. See generally, Polanyi, *Personal Knowledge*.

answer, that, for any thing I knew to the contrary, it had lain there forever; nor would it perhaps be very easy to show the absurdity of this answer. But suppose I had found a watch upon the ground, and it should be inquired how the watch happened to be in that place; I should hardly think of the answer which I had before given—that, for anything I knew, the watch might have always been there. Yet why should not this answer serve for the watch as well as for the stone? Why is it not as admissible in the second case, as in the first? For this reason, and for no other, viz. that, when we come to inspect the watch, we perceive (what we could not discover in the stone) that its several parts are framed and put together for a purpose, e.g. that they are so formed and adjusted as to produce motion, and that motion so regulated as to point out the hour of the day; that if the 2 different parts had been differently shaped from what they are, of a different size from what they are, or placed after any other manner, or in any other order, than that in which they are placed, either no motion at all would have been carried on in the machine, or none which would have answered the use that is now served by it. To reckon up a few, if I stumbled on a stone and asked how it came to be there, it would be difficult to show that the answer, it has lain there forever is absurd. Yet this is not true if the stone were to be a watch."[20]

This argument continues on at some length, but enough of it has been reproduced here so as to provide readers with the flavor of it all. A variety of teleological arguments continue to be popular today, and in what follows below are a couple of them currently in prominence.

The first one that merits mention—one that I will label "cosmic teleology"—comes from Gerald L. Schroeder, PhD., an MIT trained physicist who admits that he started early on as an agnostic, but has reacquainted himself with his Jewish roots in recent years, and now favors a anthropic approach. In this context he notes that, "the existence of life on earth appears to be surrounded with improbabilities."[21] He includes in his thinking the reigning scientific theory that the universe as we know it looks to have started with a singularity that has come to be termed the *big bang*, the event that would explain a universe that appears to be in the process of expanding. From a limited human perspective, when comparing that initial event with the cosmos in its present configuration—including the existence of life—it does seem both remarkable and statistically unexpected. To underscore the improbabilities Schroeder notes that from a cursory study of cosmology that:

20. William Paley, *Natural Theology*, chs. 1–3.
21. See Schoeder, Gerald L. *God According to God*, 20.

We reside on a very special planet at a very special location within a very special stellar system, formed at just the right position within the right kind of galaxy. The earth's distance from the sun, for the right amount of warmth, and its mass and gravity, for the ability to retain a proper atmosphere, put us in the only habitable zone within the solar system."[22] In fact Schroeder views the outcome of the big bang to be so remarkable, that for him, a physicist, it "has all the markings of being the product of design," with the study of nature representing statistical evidence for God's intervention in the development of life.[23]

The anthropic nature of this argument grows out of the appearance that the universe evolved in such a way that it must have been designed with humans in mind. There are certain points that would go into such thinking—including the precise nature of atomic structure, such that it allowed for the formation of matter, including stars. Many physicists are quick to point out that only very minor modifications in the initial conditions would have radically altered the end result—a results that could have made life impossible.

As logically compelling as this argument is, there are many critiques of this approach. For one thing, in order to know the true probabilities for Schroeder's perspective, in simple mathematical terms it would be necessary to know with completeness the frequency with which life emerges in different parts of the universe from the big bang, but also to compare the results of many "big bang" occurrences.

Mathematician David J. Hand criticizes the anthropic idea by arguing it to be a product of the law of selection. This law observes that by looking back at what actually happened, instead of looking forward and trying to see what will happen, the probabilities can be changed from being uncertain to certain. This practice has been called postdiction, as contrasted with the word prediction.[24] The point is, a very limited, earth-based human perspective is sufficiently impoverished in the grand scheme of things such that an understanding of the true probabilities is quite unrealistic, and this creates a barrier for being able to reach conclusions with complete confidence—particularly conclusions of a statistically valid nature.

Adding to the wisdom for caution in the use of statistics here, there may well be certain principles involved in the underlying order of nature that have yet to be discovered, some of which might involve some sort of self-organizing mechanisms that would modify statistical considerations. At least this seems

22. Ibid, 67.
23. Ibid, 80.
24. Hand, *The Improbability Principle*, 117–140; 218.

to be a cautionary note with at least some observed basis. Certainly for biology there is the survival instinct that would seem to propel it forward. This, in itself, is a self-organizing feature built into biology. But the biggest question for biology is that of origins—how did it all get started—though as Swinburne might assert, the simplest explanation would seem to involve some sort of purposeful design. It is this reality from which many find credibility in metaphysical beliefs that might otherwise be lacking.

A second teleological approach—what I will label "informational teleology"—is fairly complex because it gets into genetics, and will require a little space in order to develop it sufficiently. Two of the principals who have been quite active in developing this idea are Stephen Meyer and William Demski, with the argument based on three primary elements—information theory, statistics, and the nature of scientific methodology.

Meyer divides information itself into three groups—as 1) ordered, 2) complex, and 3) complex with specificity. In the case of ordered properties, the reference is to sequences that follow a set of underlying rules, algorithms, or general laws that exhibit some sort of repeating function, e.g., ABCABCABCABC. Obviously, a repetitive sequence such as this is compressible (e.g., ABC repeated 100 times, or one that can be framed as part of an exponential equation). There are some examples of ordered systems in nature that appear spontaneously when conditions are right. Examples that come to mind would be vortices, or the laws of physics that allows matter to warp space—resulting in all large cosmic bodies to be oval arranged structures. In general, ordered sequences correlate with high probability and determinism, and are represented well by the world of Newtonian physics where order and predictability are observed. There are other examples of self-organizing, including some from biology.

The second of the information groupings would be *complex information,* as distinguished from its ordered cousin, characterized by its randomness, and is not represented by any sort of compressible sequences. Information of this type was first formally considered by Claude Shannon who is regarded as one of the pioneers of information theory. He thought of information in terms of a communication channel, with focus on its carrying capacity. As such, it could be represented by a simple random distribution of information achieved by a simple alphabet soup of letters tossed into the air—with the letters coming to rest in a randomly distributed arrangement. The array will most probably be a one-of-a-kind sort, and be defined as complex due to its unique arrangement. To the extent that carrying capacity is the focus, little attention is given to decipherable meaning.

This, then, leads to the third type of information, one that Meyer references as information having the characteristic of *specified complexity.* He gets at this notion by noting the distinction between a mere improbable event (the arrangement of alphabetic letters from a random toss), and an arrangement of letters that conveys some sort of meaning. For example, consider the following alphabetical characters.

"It is a lovely sunny day"
"dkk kdksdf ererwre lee"

Both rows of letters are the same length—thus having the same carrying capacity. Both could be the product of a completely random alphabet soup of letters tossed into the air that by chance landing on the ground in the arrangements above. However, to the extent that these letters represent a completely random arrangement, it is easily recognized that they are substantively different—at least in English. Both convey information, but only one conveys functionally significant information.[25] Information scientists often say that Shannon's theory measures the "information carrying capacity"—noted above—as opposed to its "functional significance."[26] The recognition of this distinction grows as the body of information grows, and in this regard most will recognize that a Shakespeare play does not evolve by random chance.

This is the type of information that mathematician David J. Hand, refers to as part of the law of truly large numbers. It involves what has become known as Borel's law that concludes that for practical purposes, sufficiently unlikely events can be considered as impossible, and can be summed up as follows:

> Such is the sort of event which, though its impossibility may not be rationally demonstrable, is, however, so unlikely that no sensible person will hesitate to declare it actually impossible. If someone affirmed having observed such an event, we would be sure that he is deceiving us or has himself been the victim of fraud.[27]

An example of functionally significant information would include the alphabet of letters used to create words, with the specific arrangement of the letters making communication possible. In similar fashion, computers are built on the basis of binary code, and it is the specific arrangement of 0s and 1s that create the ability of the computer to act in logical and even in seemingly intelligent ways. Further, it should be noted that alphabetic and binary notation have similarities to deoxyribonucleic acid (DNA), the digital code for biological organization. DNA is composed of four nucleotide base pairs

---

25. At least if there is any meaning present in the second row of letters, it is outside of standard convention.

26. Meyer, *Signature in the Cell*, 91.

27. See citation note 112, Hand, 8.

A-T-C-G that in humans comprise a coding string that is approximately 3 billion long. It is only the specific sequencing of this code that gives biology viability, and thus, its functional significance.

Meyer concludes that in several ways, this is all very important to the conclusions that are appropriate to make. He notes that it is certainly remarkable that DNA can carry or encode information using chemical subunits that function as alphabetic characters—marveling at how it forms a part of a communication channel that can be analyzed so readily using the mathematical tools of Shannon's information theory.[28] He further notes, that "apart from the molecules comprising the gene-expression system and machinery of the cell, sequences exhibiting such specified complexity are not found anywhere in the natural—nonhuman—world. Sequences and structures exhibiting either redundant order or mere complexity are common in the chemical substrate of nature. But structures exhibiting specified complexity are completely unknown apart from DNA, RNA and proteins."[29] This is the central thrust of his argument.

Meyer makes the following point repeatedly: "human artifacts, and technology—paintings, signs, written text, spoken language, ancient hieroglyphics, integrated circuits, machine codes, computer hardware and software—exhibit specified complexity; among those, software and its encoded sequences of digital characters function in a way that most closely parallels the bases sequences of DNA."[30] From here, he goes into a lot of detail regarding the characteristics of the protein molecule and its dependence upon the specificity of amino-acid sequences of nucleotide bases on the DNA molecule. The point that he is attempting to drive home in all of this is that DNA, as an assembled string of coded information, can be reduced to a probability statement that gains importance because of its functional significance. Readers may recognize in this argument a somewhat parallel idea conveyed by William Paley, as discussed above.

Given the above discussion, it is worth noting that throughout most of human history, knowledge acquired from the biological sciences has generally been in descriptive and analog form. This is significant to the evolutionary theory, which developed from observations in nature, supported by data derived by dissecting, categorizing, comparing body plans, studying the fossil record, environmental conditions, etc. Clearly to proceed in this fashion provided a level of tentativeness that would not exist if biology operated on a more quantitative basis. With the unlocking of human knowledge of the DNA

---

28. See note 114, Meyer, 108–109.
29. Ibid, 110.
30. Ibid, 110.

code and the more recent completion of the genome map for humans and a number of other biological species, methodologies have dramatically changed biology from a qualitative science to that of a quantitative and computational science. The genetic information system is essentially a digital data recording and processing system and can now be studied on those terms.

In the pre-computer era, Claude Shannon, the father of modern information theory, proposed that communication errors could be overcome by building in sufficient redundancy and by proceeding in a discrete, linear, and digital fashion. As scientific understanding of DNA came to maturity it has become increasingly clear that DNA is at its essence, a digital code not unlike computer software code, in that it contains the specified informational sequences necessary for life. It is in this context that Hubert Yockey, a prominent physicist from the University of California, Berkeley, CA, author of the book *Information Theory, Evolution, and the Origin of Life,* states:

> Information, transcription, translation, code, redundancy, synonymous, messenger, editing, and proofreading are all appropriate terms in biology. They take their meaning from information theory and are not synonyms, metaphors, or analogies."[31]

He further notes that the existence of the biological genome and its supporting code divides living organisms from nonliving matter. On this reality, he concludes, "there is nothing in the physico-chemical world that remotely resembles reactions being determined by a sequence and codes between sequences.[32]

Science now understands that there is a genetic message recorded in the digital sequence of nucleotides in DNA, and it can appropriately be described it as the software of life that controls all biological processes. It is now understood that the sequences of nucleotides or amino acids that carry a genetic message have a precise specificity. As Yockey notes, the message in the genetic information system is segregated, linear, and digital and can be measured in bits and bytes. In the early days, biology used the only tools available, which included heavy reliance upon the comparative and descriptive. Over a period of time this has transformed into a quantitative and computational science. Yockey's conclusion is that by employing information theory, comparisons between the genetics of organisms can now be made quantitatively with the same accuracy that is typical of astronomy, physics and chemistry.[33]

---

31. Yockey, *Information Theory, Evolution, and the Origin of Life,* 6.
32. Ibid., 2.
33. Ibid., 8–11; 184.

One of the most important points for readers to grasp is that now that DNA is being sequenced for a wide spectrum of organisms ranging from bacteria and viruses, to humans and other species, the relationship of all organisms can be based strictly on the amount of similarity, or closeness with respect to DNA sequences among species. It is now possible to make comparisons, in numerical terms, between all these organisms that go well beyond mere armchair speculation.[34]

Meanwhile, the science community has been hard at work looking at possible explanations as to the origin of DNA. This search has spanned the range. This, in itself, could entail a lengthy discussion so rather than go into it here it is worthy of mention that the chance hypothesis has been looked at, as has the possibility of self-organization, as well as DNA first and RNA first approaches—among others. At this time, it is would be premature to conclude that there will ultimately be no explanations that will emerge, but such possibilities for many scientists does seem to be a long way off.

Now with this as the backdrop, it may be helpful to detail something about the methods used in studying the historical sciences such as geology, evolutionary biology, paleontology, forensic sciences, archeology, and cosmology—and all sciences that yield up answers by looking to the past. Each of them uses a different method of inquiry than does the experimental sciences such as chemistry and physics.[35] Using a cause-and-effect relationship, historical scientists have a distinctive objective—to identify the causes of past events. They calculate backwards and infer past conditions and causes from manifest effects. In other words, history is inferred from its results.

This approach is sometimes known as abductive reasoning—which seeks to determine a past or ancient cause, as opposed to establishing universal laws. Essentially, historical scientists proceed by inferring history from observed results (including unseen facts, events, or causes in the past from clues or facts in the present). This approach yields conclusions that are plausible, but not certain. The problem with this type of reasoning is that there is often more than one causation that can explain the same effect.[36]

Consequently, a method of reasoning known as "method of multiple working hypotheses" (sometimes also known as an inference to the best explanation) is used by geologists, and other historical scientists when there is

---

34. Ibid., 179.
35. See note 114, Meyer citing Stephen Jay Gould, 151.
36. First described by American philosopher and logician Charles Sanders Peirce.

more than one possible cause or hypothesis to explain the same evidence.[37] It includes causal adequacy, and retrospective causal analysis.[38]

Finally, Demski and Meyer have attempted to give statistical consideration to the probabilistic resources available by giving some thought to the time and number of trials necessary to render an event probable, using it in the calculation of the emergence of life by considering the chances of the DNA amino acid chain assembling on the basis of random chance. Mathematician William Demski developed a formula for doing this calculation, and is composed of the following components illustrated in the table below.

Probability Calculation	Event
10^{80}	Total elementary particles in the observable universe
10^{16}	Seconds since the big bang
10^{43}	Number of times that elementary particles can interact with each other per second
10^{139}	Total number of events that could have taken place in the observable universe since the origin of the big bang = a measure of the probabilistic resources of the entire observable universe
10^{164}	Probability of producing a single 150 amino acid functional proteins by chance. This number exceeds the total number of events by more than 24 orders of magnitude—more than a trillion, trillion.

The calculation of the probabilistic resources of the observable universe since the big bang was performed by multiplying # (1) by the total number of elementary particles in the observable universe (10^{80}) of elementary particles; times # (2) since there has been a limited amount of time since the big bang (10^{16} seconds), with a limited number of opportunities for any given event to occur in the entire history of the universe;" times # (3) which is the number of times per second physically speaking that an event occurs when an elementary particle does something or interacts with other elementary particles. Elementary particles can interact with each other only so many times per second since there are a limited number (at most 10^{43}).[39] By multiplying these numbers together, Demski calculated that the total number of

---

37. Developed in the late nineteenth century by geologist Thomas Chamberlain, 154.

38. Meyer cites Michael Scriven on this, 166; see also 327–344.

39. Ibid., 216.

events that could have taken place in the observable universe since its origin was 10139, vastly less than the estimated odds of producing a single random 150-amino-acid functional protein which stands at about 1 chance in 10164.

What does this mean? For Meyer it means that the only explanation for life is Intelligent Design, with him noting that "if every event in the universe over its entire history were devoted to producing combinations of amino acids of the correct length in a prebiotic soup (an extravagantly generous and even absurd assumption), the number of combinations thus produced would still represent a tiny fraction—less than 1 out of a trillion, trillion—of the total number of events needed to have a 50 percent chance of generating a functional protein—any functional protein of modest length by chance alone."[40]

While all of this may be quite interesting, neither Meyer nor Demski incorporate assumptions for the possible role of self-organization as an observed feature—not to mention other possible processes that could modify the probability calculation. Certainly unknown or poorly understood features of the natural order may have played some sort of contributory role in this equation. For this reason alone, there is an evident naivety at play by attempting to detail out statistical probabilities for the origin of the DNA code in the absence of any precise understanding of the natural processes that may have given rise to the code of life.

So as the origin of life is pondered, it is important to recognize that Yockey could well be right—such understandings may forever remain beyond scientific reach, perhaps forever relegated to tacit knowing that is the domain of sacred texts that speak of God as creator.[41] The most generous mathematical observation must surely be that based on current knowledge of the data, the odds of life spontaneously blossoming forth seem statistically remote.

To summarize natural theology, then, the observation should be made that it comes in a variety of forms, each making a contribution that in general are not mutually exclusive. In many ways it is arguable that they are supportive of each other in furtherance of the master narrative of a creator God at work in the universe.

---

40. Ibid., 218; as a response to Meyers, physicist Mailen Kootsey notes the following: "The numbers generated by this line of reasoning sound impressive to the lay person, but they are regarded as nonsense by the scientific community. The reason is this: The Dembski numbers represent the probability that a DNA molecule is formed by all the individual atoms in the protein coming together from their random motions ALL AT ONCE! No scientist would think to suggest this as a possible origin for a complex protein or for DNA. All the complexity observed in biology is generated through a multi-step process, so that is the kind of mechanism scientists think about when looking for a natural origin of life."

41. See for example, Karen Armstrong, *The History of God*, wherein she discusses at length the history of human conceptions of God. Clearly the spectrum is quite large.

Agnostic Argument

In light of the ideas put forward in this chapter there are a couple of things that seem clear. First of all, it is evident that there are a variety of ways in which to make attractive arguments in favor of a creator God, yet it is also clear that there is no way to make an epistemological argument that is compelling beyond all doubt, since a knowledge of God in any explicit sense is unattainable. This has led many to the doorstep of agnosticism. Its more extreme version is "atheism," but it is difficult to see how any true attempt at objectivity could arrive at such a position since generally it asserts an understanding of reality that claims a level of assurance that is not far removed from that of the critique sometimes leveled at Christian fundamentalists who also seem to have it all figured out.

The far more defensible position is agnosticism, either in a hard or soft form, which simply recognizes the realities of human existence for any sort of definitive metaphysical knowledge. There is a certain humility represented in such approaches and certainly the softer version can manifest a more pragmatic and modest avenue to faith.

Concluding Observations

Even though this chapter may be a reminder of the seeming hiddenness of God, a number of the above paradigms can help clarify the line between religious faith and knowledge that is either sense-based or acquired through reason. Some of these approaches overlap, and some seem more facilitative than others in advancing the credibility of belief. So, as this chapter winds to as close it is probably worthwhile to summarize some of the takeaway points. First, there is no basis in reason, or data that can prove the existence of God. Second, there are quite a number of arguments that seem to support the possibility of God. For those who are attracted to this possibility, these insights will likely be deemed as helpful. Third, there is evident tension between religious beliefs that are not carefully examined, and a more structured approach that seeks evidence. Fourth, for those who see value in having a more systematic approach to belief, there exists a wide spectrum of possibilities that can either provide a pathway for strengthening belief, or that can result in confusion given the spectrum of methodologies and approaches.

With this latter point in mind, it occurs to me that some may find these varied approaches to be of value in creating a personalized and thoughtful approach in thinking about God. For one, few will quibble with the reality that all human thought begins at the pre-logical and a-critical. It therefore becomes necessary to develop a set of presuppositions—in other words those assumptions that represent a starting point of thought, and from which all thinking is dependent. It is from this framework that Alvin Plantinga's claim that belief must be foundational can be considered. Yet, it is hard to miss the fact that *belief* alone, is subjective. The pragmatic result exhibits a wide variety of values and belief systems that often are in conflict with other systems of belief. It can therefore be accommodating to add additional steps to any analysis. In this regard, causality can be helpful in that human experience conveys to the observer a cause-and-effect world, which can lead to the quest to find the initial cause of it all. Once again, causation cannot produce any explicit results, yet it does leave open the possibility of divine causality.

Finally, some of the natural theology approaches can play a supporting role, including the use of Bayes theorem to achieve a heightened probability for the God hypothesis. Not to be overlooked are also some of the design arguments—two in particular. Some recognize the anthropic reality of human existence against the backdrop of the apparent mathematical odds, and find inspiration for belief, even as science with the help of modern magnification is discovering many exoplanets in the so-called goldilocks zone. Still others look to the information rich nature of life, its complexity with specificity, and recognize the apparent statistical mountain that must be climbed to overcome the seeming odds, short of discovery of some heretofore-unrecognized mechanism(s). Whatever conclusions are drawn on all these approaches they cannot be definitive, yet to incorporate some of them into the matrix of thinking can add substance to a master narrative that includes theistic beliefs, and do so in a way that would not otherwise be available.

So, in the end, none of these approaches can move knowledge of God to an explicit level, yet in certain ways many of these ideas can be helpful for those who desire to live life with a master narrative that includes the possibility of God, along with the meaning and purpose that it can add to life. The goal of all these approaches is to elevate the idea of God above the plane of mere opinion, and while none are completely successful, they can provide synergistic support in that direction.

The way in which a *master narrative* is framed is significant in that it guides a person through life, and the framing will dictate the possibility of living a purposeful existence. Perhaps the most significant aspect of this narrative is that control remains with the individual. Ultimately, it is not possible to escape the tacit element, yet there are a number of avenues by which a master narrative can be put on an elevated footing by way of the evidences afforded by the senses combined with the power of reason.

CHAPTER 10

Thinking Strategically

Chapter Summary

There are a variety of ways to look at just about everything. For example, it is well known that science gives studies at increasingly smaller units of all natural phenomena in a quest for understanding. This is certainly the case in biology, where endeavors are made to reduce a system to its lowest constituent parts—cellular, sub-cellular, including DNA code—in order to understand how it all works. The same method is used in the study of the non-organic world, and has resulted in the table of elements of atomic structure, as well as the quantum world of subatomic particles. This reductionistic approach has proven extremely useful and productive, yet as significant as it has been there remains an important need to also develop an understanding of the big picture—the strategic view. In biology the strategic view would be to incorporate the body of knowledge gained from all the reductionistic projects into a matrix that can offer up some high level synthesis of understanding as to the function of the organism. So as the title suggests, this chapter considers the strategic point of view for ideas that may be held—empirically, rationally, or religiously—with there being a clear hierarchy involved in thinking that builds upon certain ideas assumed ahead of time, and referred to as presuppositions. In connection with a strategic understanding of religion, it is worth noting that author and TED Prize winner Karen Armstrong has conducted decades of research on the worlds' religions in which she says there's a universal theme that ties them all together—it is the centrality of compassion, with compassion acting as a single thread running through every major belief system. Armstrong noted that as far as can be determined, Confucius was the first person to formulate the Golden Rule—this about 500 BCE She notes that when Christ's disciples said, "Which of your teachings can we put into practice all day and every day?" he said, "Never treat others as you would not like to be treated yourself." It is her view that all the faiths have come to tell humanity this, " . . . that this is what works." What she is proposing, then, is that at a strategic level, the major theistic religions are actually quite closely linked. Certainly any chance of humanity experiencing conceptions of heaven on earth would be found at this strategic level, where humans reject the law of the jungle as an operating philosophy, choosing instead to practice the golden rule. If this one principle were universally adopted, a significant part of what most people recognize to be one of the more vexing aspect of life would evaporate. Religious people unfortunately tend to overlook this strategic worldview, giving rise to a long history of denominationalism, doctrinal wars, jihads, crusades, and wanton genocide—all in the name of God.

There are a variety of ways to look at just about everything. For example, it is well known that science gives study at increasingly smaller units of all natural phenomena in a quest for understanding. This is certainly the case in biology, where endeavors are made to reduce a system to its lowest constituent parts in order to understand how it works. The same method is used in the study of the non-organic world,[1] and has resulted in the table of elements, as well as the quantum world of subatomic particles. This reductionistic approach has proven extremely useful and productive, yet as significant as it has been there remains an important need to also develop an understanding of the big picture—the strategic view. In biology the strategic view would be to incorporate the body of knowledge gained from all the reductionistic projects into a matrix that can offer up some high level synthesis of understanding as to function of the organism.

So as the title suggests, this chapter considers the strategic point of view for ideas that may be held—with there being a clear hierarchy involved in thinking that builds upon certain ideas assumed ahead of time, and referred to as presuppositions—something that Polanyi would frame as the *tacit element*. Sometimes these notions represent volitional concepts, but they can also be part of the sub-conscience. As an example, most assume that there is a real reality out there and if it is to be accessed on any level, it will most assuredly include the five senses. People may or may not be conscious of this

---

1. The distinction between organic and non-organic has been blurred over time. Certainly that which is classed as organic gets designated as such by foundationally strategic arrangement of inorganic components.

specific presupposition, but to hold it has real consequences, given that the scientific method is built, in part, on this understanding. Should it ever be concluded that reality is a complete illusion, then science becomes largely irrelevant and superstition becomes the guiding light.

There is another beginning assumption most moderns hold, with that being that reality contains logic and order. If it doesn't, then there really is no hope. However, if this notion is correct—and the scientific method seems to confirm that it is—then with the aid of sense and reason, the human capability of discovering the nature of reality on some level is opened up. So it is that from these and other such assumptions the scientific method was born, with it developing systematic rules for inquiry.

Many of the assumptions that modern humans have adopted are key to the success that science has had. For the purpose of this discussion it would be important to consider one particular non-scientific assumption that is very far-reaching and significant—one that a great many people incorporate into their thinking. The idea being referenced here is theism—that being the belief that God exists on some level as a creative force in the universe. For devotedly religious people to think of *belief* in pre-suppositional terms may seem a little clinical, yet the fact that *God* is not accessible directly to the senses and cannot therefore be empirically proved or disproved makes it appropriate to consider on a philosophical level because this idea has the capacity to order a great deal of everything else that may be contemplated.[2]

The summation of this high-level perspective is often referred to as a worldview, or a master narrative. It frames the manner in which a person sees and interprets the world around, and as such becomes critical to all other thinking that may be engaged in. It includes the collection of beliefs that may be held about life and the universe, often based on philosophical postulates or sacred claims. If a person is interested in knowing something about someone's worldview, perhaps the best way to discover it would be to engage in an exploration of his or her presuppositions, for they are key to everything else.

Strategic role of Master Narrative for Sacred Ideas

No doubt one of the most common strategic approaches to life comes in the form of religion, but even here it comes in a variety of perspectives. Perhaps the most strident of these is what is termed *fundamentalism*. In the Christian world fundamentalism has been around for about a century,

---

2. Those who assume God to be the creator of the universe, including the earth and life on it, are able to think of scientific exploration as an exploration of the mind of God. In this sense, empiricism can be understood as accessing God through the senses.

though in its modern form, some people point to a meeting in 1978 where over 200 leading Christian conservative luminaries convened in Chicago at the International Council on Biblical Inerrancy. It was here that they crafted an influential statement on inerrancy that included many points, with one of the key points being that Scripture in its entirety is inerrant—the general idea being that God superintended the endeavor, thus affording humans full assurance that no error existed in the original autographs.[3] This statement should probably be seen as strategic at least procedurally in that the way in which inspiration is understood colors everything that follows. While the 1978 statement was likely drafted by sincere and honorable individuals, those aware of the some of the issues—issues that have been briefly considered in chapter 5—have little choice but to conclude that this statement represents a vast overreach, elevating Scripture to an idolatrous pedestal since there is a great deal of evidence that the authors of Scripture were fallible humans, and that what they wrote was not inerrant.[4]

The other end of the spectrum is to see Scripture as the human attempt to frame an understanding of God. In its most extreme form it would deny that it contains any divine influence.

Between these two polar ends stands a middle course held by those who clearly see the human imprint on Scripture, yet one that would hold out the possibility that it also contains elements of the divine. Certainly, this latter position is necessary for anyone who contemplates the existence of a loving deity, as in the face of the difficult existential reality that humans face on a daily basis it is surely not conceivable that a loving creator God would exist yet fail to communicate with humanity. As with all general categories, this latter one can come in a variety of flavors in terms of what is meant by the word *inspired*. Certainly a lot of great literature is deemed inspired, but people who hold religious commitments would include the idea of divine influence in some form or fashion, while acknowledging that a lot of what is contained within its pages is of a distinctly human nature.

So at one extreme, then, are those who face the issues that grow out of the general umbrella of documents that many perceive to be the very word of God, choosing to ignore the evidence of a reality that Scripture is far from immaculate in its content and provenance.[5] Those coming from a

---

3. http://www.bible-researcher.com/chicago1.html.

4. See note 93 with citation for Stark. Here he discusses in detail some of the problems of inerrancy. But for starters, what possible evidence could validate the claim of inerrancy? It seems to simply be an unprovable assertion, but also one that appears fatally flawed for some of the reasons discussed earlier.

5. See note 40; and other numerous literary works of Bart Ehrman who transitioned away from fundamentalist Christianity due to the problems he discovered in Scripture.

perspective where inerrancy stands at the center of their faith, to consider some lesser standard would be faith shattering. Yet because of the problems this idea represents, it does seem that there would be wisdom in humility on this matter, both in terms of the claims made about the original autographs as well as the certitudes brought to the conversation. On this point it may be helpful to recall the discussion in chapter 1 which offers up differentiation between "knowledge" and "opinion," with the conclusion being that most of what humans think they know is not actual knowledge—but opinion. Once that basic point is understood, the possibility of being more receptive to other options can grow.

So, with this as sort of the backdrop, perhaps a modified approach would be to find a way of interpreting Scripture so as to promote the best chance of optimizing an understanding of reality. To start with it should be acknowledge that there are a variety of approaches that can be had and some of them may have the capacity for having a closer affinity with reality than others. Certainly a great deal can be distilled from cultural, literary and the historical-critical forms of analysis. Even though some of this can render up many important insights, it can also be very technical and tedious for the average person, and therefore such depth of study is generally limited to the scholars. For this reason, it does seem that a strategic approach is a much more accessible method that can go a long way in forging healthy interactions with a sacred text.

Professionally, I am involved with organizational development, statistical methods, strategic planning, and budgets. A strategic process tends to view things from the big picture perspective—the metaphorical 30,000-foot level—offered by mission statements, or statements of organizational purpose from which flow goals and objectives.

In thinking about some of the big questions of life, it occurs to me that perhaps it is the large picture represented in a sacred text that conveys up the most productive view—offering the potential for key insights that might be hard to see if overly focused on the details. This is not to suggest the detail as unimportant, but merely the practical acknowledgement that it is easy to get lost in the detail. It is the big picture that represents the broadest input to an individual's information grid. Such a reading searches for the key overriding principles being articulated and thereby attempts to capture its fundamental essence—something that can fit well with systematic theology.

So, what is the big picture of Scriptures that a strategic reading will provide, and how does a reader go about capturing it?

If a little exercise were to be initiated designed to summarize the entire content of Scripture in the space of a paragraph, what would be said? If it were to be shortened even further to a phrase, what would be said? If readers can figure this out—strategic thinking is taking place. As readers think about this, it is certainly worth noting the trajectory of Scripture that ultimately centers on a God who meets humanity in person, providing a window into the nature and character of the divine. So thinking strategically, one possibility that comes to mind in summing up Scripture in a phrase is, "God is benevolent and loving," with the takeaway being that this represents a window on reality at a fundamental level. If this were to be a correct understanding, then it would seem that perhaps humans were designed to optimize life by proceeding in like fashion. Obviously, there could be some debate regarding the overriding strategic essence, but likely most who are familiar with Scripture will come up with similar strategic interpretations.

When approached in this manner, it is possible to proffer critique on details that appear to fall outside of a more strategic understanding. It becomes possible, for example, to see the Gospel message of Agape love as normative, given an understanding of it representing a form of intimate divine disclosure. From this position, it is then possible to critique passages that indicates God's sponsorship of genocide—even though some of the Torah writers apparently believed that he did. By developing a strategic essence, the focus remains on those points that are central to the overall narrative, verses those that are either peripheral in importance or distorting in some way.

To the extent that one puts focus on the metaphorical trees rather than on the forest as a whole, it would be fair to assume that there will be an increased risk of running astray down a variety of rabbit trails. But by focusing strategically—at the metaphorical 30,000-foot level—it is possible to discover that some controversies simply disappear. In other cases science is capable of shedding light on some passages, with a faith community giving recognition to the fact that God's book of nature will be ignored at some cost. This is factually knowable from church history, which has already had some experience with this involving its treatment of Galileo. It did not turn out very well for those tacking against the data of nature. This is one reason why many of the most important insights from the sacred realm—manifested by way of private, rather than public evidence—remains an ongoing cautionary tale.

The Magisterium of Worldviews

Just as Scripture can be read for its strategic essence, science also operates strategically by formulating hypotheses and theories (the big picture of data). A strategic view in this arena is accomplished by empirical endeavors that have been advanced in no small part due to the discovery of the importance of the micro world—the configuration of atoms that compose the elements, subatomic structures of the quantum world, viruses, bacteria, genes, and DNA, among others. From these very small units the sciences develop models that attempt to explain functions that, then, provide a better understanding of matter and systems at a larger scale. The strategic goal of observation is to create a rational explanation for natural phenomenon, and stands in contrast to the private sources of revelation in getting at reality. The many successes of science are enhanced by way of its evidence being of a public nature; its power resting in its explanatory capacity culminating in peer review.

The obvious question emerges as to whether science and religion are two different beasts or whether it has some commonality? For sure, science is descriptive, and tends to limit its overall narrative and methods to sense based analysis—though there are exceptions as will be discussed below. On the other hand, the domain of the sacred tends to be more about purpose and meaning—though, this to, has exceptions. It is the murky middle where science and religion tend to get tangled up, and this seems to be the basis for most of the tensions that exist.

Most religious ideas emerged well before modern science at a time when ancient mythologies spoke of an unseen reality that governed the apparent whims of the natural world. But it also did something more in that it proposed that life was the handiwork of God, vesting it with purpose related to eternal concepts of life and living. Even to this day many still have a nagging suspicion—yes, even compelling commitments—to the idea that there is a tangible reality that exists beyond science, and this has all come by way of private evidence that is sometimes referred to as *revelation*. The strength of private/revealed evidence rests in part on the fact that it purports to offer up some basis for existential purpose—this in contrast with sense-based data that can make no such claim. For meaning seekers this represents the fundamental weakness of science and the strength of private evidence, though, as I have documented throughout this book, revealed source material often comes with its own set of problems.

Yet, in spite of such problems, it is noteworthy that one of the most common master narratives around tends to provide for a purposeful life by way of a sacred text. In fact, purpose can be found in the very first chapter of Genesis where "God as creator" strategically vests life with a basis for meaning found in something that extends beyond human temporality. It also frames the essence of the entire first chapter of Genesis, with no essential need to impute interpretations that would put it in an adversarial relationship with the findings of science. The importance of proceeding in this more thoughtful way comes by recognizing the role science plays in piecing together evidences of the past. There simply is no need to hold rigidly to sacred details that push the biblical narrative into a collision with science, given science's ability to effectively stand on its own two feet with a proven track record in coming to a reasonable proximity of how many aspects of the natural world function. Its tangible results stand as witness to the folly of proceeding along adversarial lines.

There is certainly nothing wrong in mining the details of the Scriptures, but when conflicts emerge between an understanding of a sacred-sourced belief and a scientific insight, it will be necessary to find a process by which the conflict can be mitigated. The process itself will be subject to the authority accorded each magisterium, with the process to be guided by an understanding that there is but one reality.

Those who proceed by way of *sola scriptura* will likely find occasions when it will be necessary to diminish the value of the senses, and based on everything that has been said about the formal pursuit of knowledge, the only conclusion that can be drawn is that this is very dangerous to the integrity of ideas derived exclusively in this fashion. On the other hand, those proceeding by way of a standard epistemological program will likely recognize that on some occasions revealed source data cannot compete with the senses. Proceeding along either track to the exclusion of the other will tend to dictate the conclusions that are achievable.

The middle ground would be one that attempts to maintain respect for both magisterium. But what does that look like?

First of all, those who assume a loving God exists will likely also assume it reasonable that he has communicated with humanity on some level, for it is simply unfathomable that a benevolent creator could observe the human plight without providing some sort of roadmap to the cosmic order. Thus, in spite of all the issues represented in Scripture there are many who stake their lives on the prospect that something of ultimate reality can be found within its pages. People of this disposition can hold a high view of both science and Scripture by seeking out a synthesis approach that attempts to bridge gaps in how these understandings play out. The search for a synthesis understanding between magisterium can proceed under the premise that the ultimate goal is to discover the reality, irrespective of a public or private divide. Having a solid grasp of the reality is the goal and any gaps in understanding between public vs. private methodologies suggests that one or the other has an impoverished hold on reality—perhaps both.

So even as science and religion primarily address different concerns, it has also been noted that there are occasions involving overlap, and at these points of convergence where science has increasingly filled in the details, some of the traditional interpretations of revealed understandings have been called into question. At such junctures, I would propose that religious interpretations can and should take the following steps:

First, perhaps the threshold question is whether the conflict is of a descriptive nature or whether it goes to purpose and meaning. If it is a descriptive program, then clearly science should be given space to do its work. After all, while a tacit element is associated with *descriptive* reality that come from empiricism, revealed wisdom is much more directly dependent upon assumptions.

Second, when discrepant understandings emerge, it is important to remember the epistemological priority that is governed by the senses, and assuming the issue does not breach the purpose level, there is wisdom in backing away from dogmatic assertions that overlap with a scientifically validated idea. For many people of faith this is a hard thing to do for it often runs against tradition, and it also can compromise the certainty that is sought in life. However, the wisdom in proceeding along this course comes by recognizing that it is not necessary to throw in the towel on a sacred tradition, but merely to afford science the room to finish its work before cementing traditional ideas in ways that may be problematic.

Third, in the midst of conflict, the sacred realm can recapture its relevance by reexamining its strategic essence—for example, by perhaps recognizing that a sacred idea is not *purpose* centered, and should therefore find a way of accommodating a scientific idea that may initially be perceived as intruding upon the sacred realm. This would at least provide a respectful means by which to peacefully coexistence.

Sensitivity regarding such conflicts should always be based on the premise that there is but one reality—not simply a science reality and a religious reality—or separate magisterium. As such, this concept would seek ways of uniting non-sensory and sensory ideas where there is a nexus between these two methodological approaches. The framework just laid out must recognize the essence of one reality as meaning that both public and private sources can be taken seriously by a person of faith, doing so in a manner that respects all legitimate pathways to knowledge. In the event of unresolved conflict, the mere requirement is to simply live with recognition of the ambiguities in play, it being intellectually preferable to disregarding one source or the other.

Non-overlapping Magisteria

Stephen Jay Gould, the well-known paleontologist, proposed what he terms, "non-overlapping magisteria (NOMA)." By use of this term he was suggesting that science and religion each have a "legitimate magisterium, or domain of teaching authority,"[6] and that these domains do not overlap. He suggests that this is "a sound position of general consensus established by long struggle among people of goodwill in both magisterium."[7]

Because of the "one reality" issue, my early reaction was to reject NOMA out of hand. However, my thinking has evolved on this because it does seem that there can be a helpful way in which to use this notion. It works well when the domains of science and religion are distinguished along descriptive verses prescriptive designations respectively. As such the sciences rely primarily upon empiricism, while the sacred realm appeals primarily to revealed sources. But with this concession in mind it still seems that the NOMA concept does not work so well when the two magisteria have overlapping issues that erupt into conflict. In what follows are some of the issues that might lead one to seek a path apart from strictly "magisteria thinking":

1. If both magisteria, as defined by Gould are attempting to understand and articulate reality in the most accurate way possible, then it would seem necessary to acknowledge that there is but one "reality," and in this philosophical sense there cannot be completely separate magisteria. Reality is what it is, and is there to be discovered—period.

2. To acknowledge this first principle creates the context to search for ways to possibly resolve conflicts that may surface—including questions of process that may have given rise to conflict.

3. It is helpful to recognize the critical role of tacit knowing as applied to all theories as found at the deepest levels of a master narrative. It should probably be pointed out that even those committed to methodological naturalism as applied to the sciences generally are unable to avoid the tacit realm, as it has played a critical role in advancing all human understanding of reality. First, the tacit element is required in making some initial assumptions about the nature of reality before any observations are made, and of course there is a tacit component attached to all hypotheses in that they offer up an explanatory narrative regarding observations for which there may not otherwise be an explanation. Physicists in particular are quite comfortable in exploring reality through hypotheses—running from speculation about a

---

6. See Gould, Stephen J. and Jonathan Cape. *Leonardo's Mountain of Clams and the Diet of Worms*, 274.

7. See generally, Gould, Stephen J. *Rocks of Ages: Science and Religion in the Fullness of Life*.

possible multiverse, to quantum string theory, and beyond. All of this requires inference.

4. Both theistic and scientific notions can spring from the same master narrative that creates a framework of understanding, and neither needs to necessarily be in conflict with the other. Though theistic notions are not generally thought of as being within the domain of "science" per se, they can be acknowledge hypothetically as being of the philosophical variety that may provide a context by which to understand reality in ways that cannot be fully tested through any scientific meaning of the word. In the case of theism, what is available are the variety of arguments that are supportive of theism, as well as private evidence that may be brought into the equation. Nevertheless, theism must be very much about the business of being in touch with reality on some level otherwise it becomes nothing more than superstition. Those who may assume the notion that there is no reality apart from empiricism will ultimately be forced to acknowledge that, if consistent, then science itself is misguided in relying upon hypothetical and theoretical pursuits, as they are rational, and not empirical per se. So, while NOMA may be a useful concept in scientific pursuits, or in certain theological exercises, it is also important to note that at a philosophical level it can become misguided if pursued to rigorously. This conclusion can be reached given the difficulty in identifying how scientific hypotheses that have no prospect of ever achieving sensate fulfillment, is qualitatively distinct from theistic suppositions. A further critical point would also be that because sensory data is so powerful, on those occasions when the two magisteria differ on some interpretation of reality there are strong reasons for sectarian interpreters to maintain a modicum of humility so as to appropriately entertain the possibility that private evidence or sacred texts may not be correctly interpreted, or may simply be wrong.[8]

5. The aspect of NOMA that has redeeming value has to do with distinguishing roles. Revealed knowledge clearly plays a role not available to the sensate pursuits of science, it having to do with issues of meaning. It is on these very important concerns to humans that it becomes necessary to look outside the sciences if a worldview is to be vested with meaning. The common source for this is religion. Alternatively, science is descriptive by nature and as such makes no value judgments on matters dealing with meaning. So in this sense science and theism

---

8 This latter point should not be too difficult to accept unless an inerrant view of private evidence (revelation) is held.

each have a separate realm of concern. It is the region of overlap where the magisteria concept breaks down.

6. Finally, perhaps the easiest way to navigate through conflicts that emerge, is to give priority to private sources of knowledge at the most strategic level. For example, the most strategic level of Genesis One is that God is creator. With this strategic essence in place, people of faith can allow science to speak to contemporary people to the extent that it is capable of doing so. This may require that much of the chapter to be read as metaphor, it still holding importance because of its pointing function to the reality beyond the metaphor.

Making the most of Tacit Knowledge

Recently I had a conversation with a former colleague in which I was describing to her some of the contours of this literary project. She happens to be Buddhist, and while Buddhism does not directly address theism, she has personally concluded against it. She rather dismissively asserted God's nonexistence as if this was an idea that could somehow be affirmatively divined—pardon the pun. I did not think the issue worth pursuing at that moment, however, if there be any point at all to this book, it would be that when categorical statements are made, the question must be entertained— "what exactly is it based on?" What is the thinking methodology being used that leads to such a defining conclusion? Many hypotheses have no prospect of ever achieving absolute resolution, and certainly theism fits into that category, yet the evidence that would be supportive of this idea can be discussed, as can the evidence that might call it all into question. Likewise, the faith that bridges some level of knowability can also be considered. In all such cases, there is really very little basis for making a definitive conclusion. The way forward then is at a tacit level of personal commitment to proceed down one path or the other—mindful of the vulnerabilities that exist. As an acquaintance of mine said recently, "I believe very little, but I commit to a lot." He was here acknowledging the place where all thoughtful theist must be.

A more extreme version of this existential reality is to take the ambiguities of religious belief and turn it into an operating presupposition that does not need evidentiary support one way or the other—it is simply part of a master narrative that provides meaning to existence. When Alvin Plantinga argues for God's existence—by asserting an inner feeling, this would be his implicit point. If that is the approach, well, fair enough. It certainly is as defensible as a categorical claim that "God does not exist." Yet to be clear, neither of these extremes will likely be entirely satisfying for most thinking people, and it simply might be better to admit that reality presents many uncertainties about which intellectual confidence can be compromised.

The middle course, then, between philosophical naturalism on the one hand and a superstitious course that ignores data, leaves open the possibility of theism, but at the same time chooses to follow sensate data as far as it will lead. There is a natural spectrum for those taking this latter approach, with some choosing to remain agnostic on the question of theism, while others will be inclined to incorporate theism into a master-narrative by making certain commitments in the face of incomplete information. Within this spectrum falls methodological naturalism which attempts to take empirical data as far as it can go, without offering up any need to make assessments as to whether God does or does not exist—generally focusing on the naturalistic processes that science is capable of studying. To proceed in this way allows for refraining from judgments about the nature of ultimate reality on science related issues, but does not preclude having a personal perspective. Thus, while anything outside of a sense-based investigation is not classified as "science" per se, at the same time it is inclined to leave open the nature of reality that may or may not exist beyond the capacity of human perception—either now and/or in the future.

Summary Thoughts

In connection with a strategic understanding of religion it is worth noting that author and TED Prize winner Karen Armstrong has conducted decades of research on the worlds' religions in which she says there's a universal theme that ties them all together—it is the centrality of compassion, with compassion acting as a single thread running through every major belief system. Armstrong noted that as far as can be determined, Confucius was the first person to formulate the *Golden Rule*—this about 500 years before Christ. She went on to note that when Christ's disciples said, "Which of your teachings can we put into practice all day and every day?" he said, "Never treat others as you would not like to be treated yourself." It is her view that all the faiths have come to tell humanity this, " . . . that this is what works." She goes on to state that, "By looking into your heart and discovering what gives you pain—and then refusing to ever inflict that pain on someone else" . . . we "achieve new capacities of mind and heart." Putting the Golden Rule "into practice every day," she says, "is what matters."

What she is proposing, then, is that at a strategic level, the major theistic religions are actually quite closely linked. Certainly any chance of humanity experiencing conceptions of heaven on earth would be found at this strategic level, where humans reject the law of the jungle as an operating philosophy, choosing instead to practice this Rule. If this one principle were universally adopted, a significant part of what most people recognize to be a troubling aspect of life would evaporate.[9] Religious people unfortunately tend to overlook this strategic worldview, giving rise to a long history of denominationalism, doctrinal wars, jihads, crusades, and wanton genocide—all in the name of God.

In the next chapter the Golden Rule aspect of life will be put to the test as consideration is given to the tribal instincts of humanity that far too often tends to divide and then deteriorate from there into destructive wars.

---

9. See generally, www.ted.com/talks/karen_armstrong_let_s_revive_the_golden_rule.html: see also http://www.youtube.com/watch?v=8idmgp4icq4.

Chapter 11

Surmounting Tribal Pathology

Chapter Summary

Humans are tribal by nature, serving as part of a natural social mechanism in coping with a hostile and indifferent world. It often manifests in quite unconscious ways when people think in "us" vs. "them" terms. Some of the tribal categories that come to mind are the following: family, ethnicity, religious, doctrinal, political, education, social class, ideology, and nationality. Tribes, when considered as general categories, are frequently benign, yet they often manifest in pathological ways and are capable of having toxic effects on those inside the tribe that then ripples out in negative ways to those outside the tribe. Sometimes these effects are on full display when tribes critique in a judgmental sort of way, those outside the tribe. It is not uncommon for tribes to sometimes attempt imposition of their values on those outside the tribe, and this can occasionally lead to wars—unnecessary wars. Most astute readers will be familiar with such goings-on, which evidences itself by talking heads in the mass media hammering away on issues—both mundane and significant—couched in "us vs. them" terms. Such communications are often designed to inflame prejudicial passions on one side or the other, and the toll it exacts can result in a coarsening of the social fabric that then ripples out into an increasingly dysfunctional society where one tribe is pitted against another. Many who identify with an ecclesial tribe embrace its worldview uncritically; other tribal members may sometimes offer a pretense of objectivity by giving a veneer of critical assessment to understandings, yet do so with varying degrees of self-deception regarding the veracity and strength of the embraced dogma. In chapter 2 discussion was had as to some of the ways in which self-deception can occur including: treating a master narrative in uncritical and idolatrous ways; making something out of nothing; making too much out of too little; seeing what one wants to see; and other mistakes of methodology. In short, the human drive for certainty often clouds judgment. In his monumental work, The History of Civilization, Will Durant reflects on this history by noting, "... we can only mourn over the absurdities for which men have died." The absurdities are represented in the many holy wars that have been fought throughout recorded history over matters that, at base, fall within the scope of tacit—not explicit—knowledge. Those who understand the tacit nature of the sacred, as important as it may be to the knowing process, will be less prone to excessive behavior. Even though much of this dark past may seem rather quaint and quite irrelevant to contemporary life, and even though it is easy to think that the present order has matured beyond the reach of historic repetition—the current trend towards pathological forms of tribalism should be a reminder that dangers still lurk. In particular, once an issue is reduced to good vs. evil, then the ends often justify the means, frequently spilling out in dangerous and harmful ways. The proposal is made that the way in which the destructive form of tribalism is overcome, is to alter the focus from tribalism to that of common bonds. From a distinctly Christian perspective the core message is not so much tribal as it is one that holds out charity towards the other to lift them up. It seems that the point of primary concerns include that of looking after the economically disadvantaged, the sick, and the stranger. These are the principles that appear to assume importance. There is no other practical aspect of the Gospel commission that supersede this point and so obviously this would have implications for those who believe that denominationalism, doctrinarism, or some other "ism" is of paramount importance. The final exam seems to boil down to the effort committed to advancing those common bonds that bind humanity together in uplifting ways.

On January 7, 2015, masked gunmen entered the offices of Charlie Hebdo in Paris, France, killing twelve people. By all accounts it was satirical cartoons lampooning the prophet Mohammed that inspired this particular episode. Subsequent to this was a much more brutal attach in Paris which left around 130 dead. While in each of these cases the perpetrators invoked Islamic grievance as the basis for these atrocities, they are not alone in preaching a theology of hate that then evolves in violent ways. In the United States, for example, there are a number of groups that express overt racial biases, including: The Christian Identity movements, Aryan Nation, and Ku Klux Klan. There are also groupings that occur around theology as well:

this would include a number of right to life groups, Christian Reconstructionism that seeks to impose theocracy, and others. The common thread between all such religious extremists seems to be a desire to purify the world for God by elevating one group at the expense of another.

Tribes

So, with this introduction, perhaps it is worth noting that humans are tribal by nature, serving as part of a natural social mechanism in coping with a hostile and indifferent world. It often manifests in quite unconscious ways when people think in "us" vs. "them" terms. Some of the tribal categories that come to mind are the following:

1. Family
2. Ethnic
3. Religion
4. Denomination
5. Doctrinal
6. Political
7. Education
8. Social Class
9. Ideology
10. Nationality

Tribes, when considered as general categories, are frequently benign, yet they often manifest in pathological ways and are capable of having toxic effects on those inside the tribe that then ripples out in negative ways. Sometimes these effects are on full display when tribes critique in a judgmental sort of way, those outside the tribe. It is not uncommon for tribes to sometimes attempt imposition of their values, and this can occasionally lead to wars—unnecessary wars. Most astute readers will be familiar with such goings-on, which evidences itself by talking heads in the mass media hammering away on issues—both mundane and significant—couched in "us vs. them" terms. Such communications are often designed to inflame prejudicial passions on one side or the other, and the toll it exacts can result in a coarsening of the social fabric that then ripples out into an increasingly dysfunctional society where one tribe is pitted against another.

While any given tribal category referenced above can exhibit pathology, in the context of this particular conversation I would like to focus primarily upon the ecclesial community, which as a category is composed of a great many denominational tribes. Unfortunately, history provides abundant evidence of the pathology that sectarianism generates, replete with schisms, crusades, inquisitions and jihads. This is particularly ironic given that ecclesial tribes deal with issues that are often a lot more nebulous than those taken up by many of the other tribal categories and as such would logically seem to be less prone to pathology. In fact, history has tended to suggest otherwise. For one thing, the realm of the nebulous is ripe for multiple understandings and interpretation. This, then, has resulted in the evolution of a multiplicity of faith communities around which a tribal identities form. In turn, conflicts that sometimes turn violent have all-too-often been the end product.

Many who identify with an ecclesial tribe embrace its worldview uncritically. Tribal members may sometimes offer a pretense of objectivity by giving a veneer of critical assessment to understandings, yet do so with varying degrees of self-deception regarding the veracity and strength of the embraced dogma. In chapter 2 discussion was had as to some of the ways in which self-deception can occur including: treating a master narrative, such as church dogma, in uncritical and idolatrous ways; making something out of nothing; making too much out of too little; seeing what one wants to see; and other mistakes of methodology. In short, the human drive for certainty often clouds judgment.

When Pathology Takes Over

In situations where narrative (dogma) is more important than data, it is possible to lose touch with reality, and thereby become irrelevant to the larger world. This can be particularly the case when points of dogma overlap with the sciences. Too often the tendency is to construct the helpful data around the dogma, but then to go on and ignore or discredit the unhelpful data. This is common practice in many realms of discourse, but seems to be particularly prevalent within ecclesial communities. When narrative is built on a perceived inerrant interpretation of Scripture there simply is little room for any further discussion. More significantly, there can be no assurance of being on the right track short of taking seriously all relevant outside data having the capacity to modify the dogma narrative. This is a genuine problem within a segment of the ecclesial community and I know individuals in whom narrative is sacrosanct and thus where nothing is allowed to dislodge it—not a scientific consensus, nor "worldly" data, nor anything of historical relevance that might suggest a need for caution—Galileo notwithstanding.

In fairness to those who most observers might describe as engaged in idolatrous parochialism, it is important to distinguish between two types of tacit knowing. First, would be the observation that all thinking begins with a set of presuppositions, and it is from the "tacit" that presuppositions develop and narrative forms. In this sense, the science narrative is no different from a faith-based narrative or one full of superstitious ideas. Many in the faith community are quick to make this observation, but if the analysis stops here a critical mistake is made. The mistake comes in failing to recognize what comes next, principally the fact that any presupposition deemed credible must be accountable to real world data where such data may be available. So it is in this second, more discriminating sense that if the data suggests the possibility of a narrative being off base it may be worth taking another look at the presuppositions involved in terms of either being over-broad or simply wrong. After all, the ultimate goal must be to chips away at all forms of superstition and the human fog of ignorance by looking for data that can explain the observed reality.

There are clear examples of the ecclesial community having sometimes taken wrong turns—particularly on matters that overlap with the sciences. Perhaps the most vivid example historically was referenced above, and that is the idea of geocentricity, which at one time possessed scientific currency. The Church adopted this idea because it meshed well with the overall theological narrative that could be argued with the help of a few key texts, namely that humans were cosmically special as evidenced by the heavens revolving around earth. Yet as the data accumulated over time it became increasing clear that this idea was fallacious. As the data came in, the science community made adjustments to their own narrative so as to align it with real world findings. Ironically such nimbleness is often criticized in many parts of the faith community—where strength is argued to be a weakness. Such talk is usually framed with the idea of the superiority of religious dogma because of its perceived unchanging character. Very often the parochial approach has been to act tenaciously so as to resist making necessary adjustments, with the arrogance of proceeding thus surfacing when credibility and relevance requires pursuing a different course. After all, credulity that is foundational to superstition must succumb to real world data if rationality is to prevail.

I would contend that in the modern world any ecclesial community that refuses to give respect to the dominant scientific consensus regarding issues that overlap with a reading of Genesis are in great peril of irrelevancy—if not at the moment, then over time as the currents of data move decisively against the parochial understanding. Even if the scientific understanding is incomplete or perhaps even incorrect, it is not possible to claim the high ground in pretense supportive of "truth" and at the same time ignore a prevailing scientific understanding. If the scientific data is simply incomplete, or is being interpreted incorrectly, science will eventually figure it out and modify its narrative accordingly. Unfortunately, most ecclesial communities are not quite so agile.

Religious communities are often known for being strong-willed and not always accountable to certifiable realities, even though history suggests that there can be wisdom in doing so. In some cases, it may be useful to reexamine parochial presuppositions when in seeming conflict with the data. Sometimes all that may be necessary is a possible narrowing of the claims otherwise made.

On controversial issues within a tribe, it is possible for a situation to become so contentious that fracture of the tribe can occur, with tribes emerging within tribes, or the most pathological and destructive manifestation—tribe against tribe. To avoid fracture there may be some value in simply giving voice to an emergent conflict, for example, between faith and science by choosing to either narrow the religious presupposition(s), or to patiently await further data (for the Christian community this would include both data from nature as well as from the sacred), recognizing that such may be necessary in order to avoid mistaken conclusions. It is possible, for example, to acknowledge God as Creator as a starting presupposition, but then recognize the early chapters of Genesis offering up a statement of metaphorical value, communicating something important without demanding that it be read on literal terms as a science essay.

In many of these situations patience and humility are essential. Unfortunately, the tendency of some is to rally around a narrative tradition, and do so dogmatically, ignoring the headwinds of emerging and countervailing data.

When objectivity is the goal, it is always best to starts with presuppositions that are framed as narrowly as the data requires. This approach has been overshadowed in recent years, socially, culturally, and politically by a narrative-first approach, and when narrative becomes the overriding concern data is subordinate to that agenda. In the sectarian setting, proceeding in this way must be understood as naive at best and disingenuous at worst. As noted above, it can become the basis for schism, and war when enough negative energy is vested in rooting out heretics, i.e., those within who are labeled as being member of a different tribe. Even where it is demonstrable that the path being taken will negatively impact the common good, history is quite clear that many times an institutional system gets blown up.

Perhaps the first sign that pathological tribalism is afoot is when institutions or commentators engage in a one-sided conversation defending the narrative, while ignoring elephant-in-the-room data that might modify the traditional narrative. As a practical matter, objectivity requires all data to be put on the table—even data that would tend to run counter to the narrative. Data, taken as a whole, is not biased—it just is. Furthermore, data does not show up and tell an observer how it should be interpreted; is often less than transparent and generally can have multiple interpretations, sometimes presenting a situation that is both complicated and murky. But anyone truly interested in "truth" will never summarily dismiss a peer-reviewed scientific consensus. Yet, for the narrative-first crowd there will seldom be any inclination to give voice to a scientific consensus or complexities of data, nor give more than lip service to balance. To the extent that conversation proceeds as though issues all come in either black and white, or as if the only options were either/or, thoughtful people will likely find it difficult to participate in the tribe.

It may be remembered that the early discussion centered on concern for tapping into reality by examining the whole process of knowing. Realistically if humanity is to make any progress, knowledge and the formal process by which it is acquired must assume a central point of concern. Such concern will necessarily spill over into all aspects of life whether it is of a scientific, political or religious nature. As knowledge becomes the point of focus, a corresponding reduction in arrogant certitude, or general partisanship can take hold, as the focus shifts to an open and enhanced understanding of reality.

In order for this transformation to take effect, it must be understood that the process of knowing is far different from the human tendency to render opinions—based on traditional understandings of Scripture—dressed up as knowledge, and I am specifically referring here to matters associated with religious dogma and belief. By now it should be relatively clear that genuine knowledge is difficult to come by, and this fact grows exponentially when moving beyond the sensate level. Yet humans are far too often incline to proceed with certitudes built on a foundation of sand. The mind is easily led astray with "belief" often getting expressed in dogmatic ways.

For many religious organizations, right belief is given high priority, and the vast number of Christian denominations highlights the fact that the quest for knowing is no small issue. The overall inability to achieve consensus can be illustrated by the tens of thousands of sectarian organizations that do not have enough in common to unite in common purpose—not to mention the 4,000-5,000 other religious belief systems, many of which are also fractured. If nothing else, this fact is surely a commentary on the human condition which manifests a hunger for the ever elusive transcendent reality that can provide a framework of meaning. The multiplicity of belief systems should probably be recognized as some indication that humans are largely walking in the dark.

For sure, Gods book of nature can be seen as conveying some understanding of the divine mind, but the inability to examine, in a sensate sort of way, the transcendent force behind it all obscures the human ability to make dogmatic pronouncements regarding it. Nevertheless, it is not uncommon for individuals to push back against the aforementioned existential anxieties by sometimes giving unwarranted certitude to a belief, even though convictions without the appropriate foundation can trick the mind into believing that life possesses more assurance than what actually is appropriate. I tend to think that examples of this can be illustrated by well-intended terms that reference both the good and the bad that unfolds in life through assertions that sometimes take the form of identifying a given situation as being a part of "a divine plan," or "divine blueprint;" or conversely asserting "God's protective hand" in a given situation, or claiming to be acting on behalf of God. Such characterizations are human attempts to infuse the mind with the strength for dealing with the many vicissitudes of life, but none of this comes without major philosophical problems as has been discussed, nor is it grounded in the senate world. It is in the same way that categorical assertions about God having created in six literal days a few thousand years ago can be seen as a dogma construct that may come with some personal satisfaction, but does so at the expense of denying real world data.

Once constructs such as those mentioned above get framed, they are often incorporated into the master narrative, and then elevated as formal or informal ecclesial dogma. Obviously when ideas such as these get elevated to an inappropriate level of authority, the next step is to vigorously defend them—and as history has shown, sometimes even in violent ways. The past—even the recent past—is a reminder that wars sometimes get fought and people die over matters believed by the tribe to have a "rightness" about it, even as history has often judged some of these issues to be extraneous or insignificant in the grand scheme of things. While religious warfare may seem rather quaint and irrelevant to contemporary life, a 9-11 event sort of brings things back into focus. It becomes a reminder that the human capacity for savagery by way of sectarian certitudes that have no sensate basis, is still alive.

The tribal mind is at work in fundamentalist impulses that surround the three great monotheistic religions—where theology is sometimes used to demarcate good and evil. This can sometimes be evidenced in hyper forms of denominationalism, creedalism, doctrinarism, and associated fanaticism. The historic way in which the world has been cleansed of heretics and infidels who are perceived to corrupt the status quo has been through shunning, excommunications, inquisitions, crusades, jihads, and other forms of "holy wars." Religion, which ostensibly seeks peace on earth, has often played a prominent role in the instigation of conflicts throughout the world, and as Karen Armstrong points out, this stain has largely been the work of the world's monotheistic religions.

In his monumental work, *The History of Civilization*, Will Durant reflects on this sordid history, noting, " . . . we can only mourn over the absurdities for which men have died." The absurdities are represented in the many holy wars that have been fought throughout recorded history over matters that, at base, fall within the scope of tacit—not explicit—knowledge. Those who understand the tacit nature of the sacred, as important as it may be to the knowing process, will be less prone to excessive behavior.

In Search of the Common Good

Even though much of this dark past may seem rather quaint and quite irrelevant to contemporary life, and even though it is easy to think that the present order has matured beyond the reach of historic repetition—the current trend towards pathological forms of tribalism should be a reminder that dangers still lurk. Jonathan Haidt, who is a social psychologist, discussed with Bill Moyers some thoughts regarding ideological pathology such as that which is so visible on many fronts these days.[1]

While tribalism is a normal human response to hang-out with those who are familiar and who share common interests, values, and purpose, it is its unbridled form that is problematic. It can be blinding to those so engaged and can lead to wars—unnecessary wars. Tribal instincts often play out through vilification and demonization, as well as by reducing issues to good vs. evil. This latter category is particularly problematic, as Haidt points out, because once an issue is reduced to good vs. evil, then the ends often justify the means, frequently spilling out in dangerous and harmful ways.

Haidt proposes that the way society (or an ecclesial community) can extricate itself from a destructive path is to alter the focus from tribalism to that of common bonds. This idea dovetails back to the discussion in chapter 8, regarding *story* and its central function, as key to living healthier and more integrated lives. For one thing, each individual controls their own *story*, with story being spun out in destructive tribal ways versus processes that support that, which binds human social groups together. While it is sometimes more comfortable to be tribal in a myopic sense, most individuals will also

---

1. http://billmoyers.com/segment/jonathan-haidt-explains-our-contentious-culture/.

recognize that long-term tribal interests are likely to be best served through a commitment to the common bonds of the tribe.

There are a great many things that create common bonds, not the least of which should be stewardship over governing institutions that have been entrusted by the generations that came before. Pivotal to how these divergent situations evolve—whether governmental, business, or religious—can be as simple as whether members of the tribe are capable of rising to their better natures or whether they will retreat to destructive tribal ghettos. If the latter is chosen, there is a pretty good chance that the future of the affected institutions will be altered in ways that may later be regretted.

From a distinctly Christian perspective the core message is not so much tribal as it is one that holds out charity towards the other to lift them up. It seems that the point of primary concerns include that of looking after the economically disadvantaged, the sick, and the stranger. These are the principles that appear to assume importance.[2] There is no other practical aspect of the Gospel commission that supersede this point and so obviously this would have implications for those who believe that denominationalism, doctrinalism, or some other "ism" is of paramount importance. The final exam seems to boil down to the effort committed to advancing those common bonds that bind humanity together in uplifting ways.

---

2. See for example, Matt 25:33–45.

CHAPTER 12

Sacred Credibility in the Twenty-first Century

Science cannot solve the ultimate mystery of nature. And that is because, in the last analysis, we ourselves are part of nature and therefore part of the mystery that we are trying to solve.

—MAX PLANCK[1]

Chapter Summary

To understand that there is a tacit component associated with all knowledge, can infer the wisdom of modesty in all truth claims—whether scientific or of a transcendental sacred nature. Recognizing this fact tends to level the playing field slightly between the ethereal and empirical, given that the overriding point made is that much of the assumed concreteness of knowledge does not in fact exist in any absolute sense. Nevertheless, even though the tacit component is always in play, the sciences do have the capacity to capture degrees of concreteness not otherwise available simply because of a direct tie to the empirical world where assumptions can be tested. More important though must be the recognition that while science and religion have some overlap, in many respects they each operate in differing realms, as science attempts to explain on the basis of observation of physical phenomenon, while the sacred addresses the transcendent and is primarily concerned about questions of purpose and meaning—things about which science has nothing to say. The success of ethereal quests must be measured in part on how connected it remains to empiricism, that method most connected to the real world and whether it elevates or destroys the human spirit. In

---

1. Planck, *Where Is Science Going?* Trans. J.S Murphy.

contemplating the uncertainties associated with the sacred paradigm, philosophers and theologians have created a rich series of logical formulations for putting theism on an enhanced foundation. At the end of the day, no one can explain with complete satisfaction how a loving God can coexist with a broken world of suffering, nor why a universe that countenances both moral good and evil should be so full of mystery. Yet in spite of the loose ends, many have found sufficient viability in the Christian master narrative so as to live in the hope that reality at a fundamental level is consistent with this narrative. It is a narrative that has inspired hope for billions of people through the millennia of time, and lives on to capture the imagination and passions of many more. It has been portrayed very effectively through the arts, through grand architecture, and through compelling music. The universal message of these great masters has been to convey the language of a transcendent hope, but a hope grounded in the real world.

In this final chapter, summary consideration will be given to ways in which religious dogma can be framed credibly so as not to violate foundational elements of human knowability that at a most fundamental level has to do with human sense and reason. Any dogma that needs to be sustained by censuring the sciences requires a mature knowledge of its method if a critique is to be credible, and must resist the urge to use shortcuts in the thinking process. In this regard, this entire discussion has been intent upon providing the tools by which readers might discern a reliable approach from those that tend to operate from ignorance, self-deception, or some level of dishonesty.

By starting with the recognition that there is a tacit component associated with knowledge, it has been argued in the foregoing pages that there is wisdom in modesty regarding all truth claims made—with this ranging from the scientific to the transcendent sacred realm of religious dogma. In all of this the tacit component has an equalizing effect to some extent. To acknowledge this gives no license to ignore or misuse avenues of knowledge that convey information inconvenient to an operative master narrative—which in sectarian terms would be summed up by way of dogma.

In navigating these conflicts one of the fundamental objectives of this book has been to develop a reliable systematic process by which a person might have relative confidence of being on the right track. In order to sufficiently develop this discussion, it has been necessary to consider a number of points. Among these was one that reflected on differing types of knowledge—absolute knowledge (definitional, quantitative), and empirical knowledge (involving the study of data, testing and observing the results, from which conclusions can be drawn based on reason). Beyond this are opinions which often masquerade as knowledge, but which are not.

Because empirical knowledge often relies upon experts, a part of the perplexity for many laypeople is the definition of an *expert* where sometimes anyone with a terminal degree in some part of science is deemed an expert on all scientific questions. From the lay point of view such individuals are indeed experts by comparison, yet if genuine knowledge is the goal then it is important to recognize that mere scientific training does not adequately equip individuals to speak with authority on subjects outside their area of formal study. In a sense, it is easy to see how this issue arises, because when a speaker talks authoritatively about matters that the listener knows little to nothing about, it can be quite easy to simply capitulate. In fact, this is such an entrenched problem it was important to layout some guidelines, further noting that even the community of experts sometimes have outliers that march by the beat of their own drum. Thus, laypeople really have no choice but to lend a great deal more credibility to ideas that have been through the peer review process of the actual experts. It is the peer-reviewed ideas that get elevated to the level of *knowledge* as discussed earlier.

Also in the mix for consideration were various mistakes of reason that often lead astray, with the overriding blunder being that of confirmation bias, it being supported by: 1) making something out of nothing, 2) making too much out of too little, and 3) simply seeing what the holder expects to see, but missing a larger and more compelling reality.

In order to adequately navigate the *process* terrain, attempts have been made to demonstrate the necessity of considering things like sources, structures, content, limits, and vulnerabilities. Many chapters have been devoted to taking all these issues into account. As such, it has been essential to delve into subject-matter regarding the sacred that some readers may have found uncomfortable. This conversation has been needed—not to tear down—but as an attempt to reset how some of these issues are approached. One thing is for sure, in spite of the fallibility of the senses they do keep humans grounded in a way that is not otherwise possible. Certainly there has been no intent to suggest that religious presuppositions as a category are inappropriate—simply the importance of understanding their epistemological limits.

Such issues sometimes come up amidst conflict where revelation (or its interpretation) is clashing with sense and reason. In such instances, it is not uncommon for those from the faith community to assert that the senses are not to be trusted. Perhaps in most cases this is motivated by efforts to elevate the authority of Scripture.

As discussed in a number of earlier chapters, I would be the first to admit that the senses do often present challenges. There are many examples of empiricism having failed on some level. Perhaps the most prominent example of this sort was the Ptolemaic understanding of a geocentric cosmology—something mentioned in an earlier chapter. In that situation the senses conveyed that the earth was at the center of everything else, and this due to the feedback received by observing the perceived arcing of the sun, moon and stars around the earth. In short, the senses led observers astray. It wasn't until Copernicus theorize a heliocentric model of cosmology in order to resolve certain anomalies—later aided by strong evidence produced by Galileo—that a whole new paradigm took hold. Simply put, the intricacies and complexities of physical reality do sometimes require modifications of previously held views, resulting in the possibility of paradigm shifts.

The point is, sense-based reality is fair game to critical examination, but so also the general category of revelation—generally the source of sacred belief. Both can be a part of any thorough-going analysis in the process of connecting to reality. In fact, there was an attempt to demonstrate that sacred sources, doctrinal constructs, and unwarranted commitments to religious traditions and institutions all come with their own set of issues and therefore efforts to diminish the senses in an effort to bolster sacred categories is little more than a fool's errand.

If earlier chapters have attempted to do anything at all, it has been to demonstrate that sacred claims come with their own set of problems—problems that many religiously inclined people seem rather reluctant to contemplate or admit. For one, it is critical to the credibility of a master narrative to recognize that just because the senses can occasionally deceive, this is no reason to assume that sacred propositions (or interpretations) are somehow beyond error—either as authored or interpreted. This is a very significant point, because in certain important ways the level of authority given to sacred writings will influence the role it plays in the formation and maintenance of a master narrative. Acceptance or rejection of sacred ideas, as well as the degree with which they are held, are generally going to be relative to what *inspired* actually means to the holder. For some a sacred text means "inerrant and infallible"—in which case the text gets elevated to a level that is beyond critique. Proceeding in this way, either by assertion or practice, tends to create a metaphorical cognitive box, making it difficult to see other possibilities even when it means an idea will be unaccountable to verifiable facts. If objectivity is to be a valued part of the equation, recognition must be given to the fact that everything possesses vulnerabilities and nothing is beyond critique—including revelation.

The importance of such reflections move to the fore on those occasions when an operative worldview collides with sense-based data, and particularly where a master narrative runs against the information conveyed by the senses. When this happens, the idea becomes unaccountable to other important forms of knowing.

Any effort to incorporate transcendent matters into a credible matrix of understanding must resist postmodernist impulses, poorly constructed processes, self-deception, and certitudes of mind, if the goal is to attempt genuine proximity to the realities. The alternative is to live in a fantasy world that is poorly connected to reality and that can often lead down a descending path into a range of tribal pathologies. This is all a part of the ontological lay of the land confronting some branches of twenty-first century religion, and while epistemology can be a great tool in navigating some of this, it does so without pristine absolutes due to the tacit component in play.

The Quest for Meaning

So, as consideration is given to the triad of *empiricism, reason* and *revelation*, recognition goes to these tools as representing the primary methodologies available in any attempt to understand the universe on meaningful terms. Within the preceding pages it has been noted that science has limited its inquiry for valid reasons to the sensate realm, embracing only the "sense" and "reason" part of this triad. Because of this limitation, some have been critical of the sciences, wanting to use revelation to push it further than it can legitimately go. Yet, in spite of the criticism its credibility is built on the successes it has had in making the world a more habitable place. Its accomplishments are on display everywhere pushing back against this critique of disdain. But success aside, it is significant to note that *meaning* per se does not transparently emerge by giving study to the material world—this being one of its major limitation. Science is innately descriptive—not prescriptive—and this is the crux of the problem for most people who seek meaning.

While the *sacred* does not emanate from the evidentiary foundations of science, the sciences do have the ability to observe the pervasive influence it has on humans as well as society, to probe its physiological manifestations within the human brain, and its universal impact throughout history. With this in mind I would like to briefly delve into the narratives that frame both naturalistic and transcendent presuppositions, and the commentary they each offer in the quest for meaning.

Turning first to naturalism, it may be useful to simply reflect on the current state of human reality, namely that in many ways contemporary life represents the best of times as science has made great strides in bringing to civilization a much-improved life—automobiles, air travel, telecommunications, including cell phones, land-lines, radios, televisions and computers, among all other forms of technology. The world of medicine has likewise made great strides, and this has included the taming and conquest of many of the great scourges such as polio, smallpox, and bacteriological borne diseases. This is the age of gene mapping, and puts humanity on the cusp of a host of genetically tailored therapies. Indeed, these so-called "best of times" have been afforded largely by science and the magical world that it has brought; things that could only have been a fairytale dream to earlier generations.

A little over one-hundred years ago, at the start of the twentieth century, the reality of a much-improved world coming online at a steady clip through scientific discovery created a lot of optimism. Over the intervening years, the pace of change has continued unabated and has, in fact, accelerated. These can be occasions to celebrate and assert that these are the best of times, yet, it is clear that much of the previous optimism from those earlier generations has evaporated. Much of the scientific outlook has been pierced by growing doubts about some of the negative consequences that have been spawned by the success of science—a nuclear and biotechnological world with proliferating weapons of mass destruction, population growth that has reached the breaking point, and an increasingly polluted world resulting in climate change. It is finally dawning on many observers of the world scene that though this may be the best of times, it is also paradoxically the worst of times.[2]

The fact that pessimism has replaced optimism would seem to be partly a function of both postmodern and sectarian influences, which increasingly seem to pursue tribal agendas that are socially destructive. For sure, it would seem that the gloomy mood that prevails is not an outgrowth of learning about the long run cosmological realities (of the sun depleting its energy), primarily because it remains such a distant event. More probably it is attributable to growing doubts about the human short-term ability to sustain life, with there now seeming to be a growing recognition that previous optimism was misplaced in the face of the proliferating human capacity for self-annihilation—including in some cases sectarian pathology. More than a few seem obsessively driven to achieve this very end.

Unfortunately, it is not just the sectarian community who see the end to life on earth. There is also a growing secular recognition that humanity

---

2. Many cosmologists and theoretical physicists do give rational thought to some of these currents. See for example Martin Rees, *Our Final Hour,* 2003; see also Fred Adams and Gregory Laughlin, *The Five Stages of the Universe*; see also Dan Falk, *In Search of Time*, 2008, specifically his chapter entitled, "All Things Must Pass."

and all other life are endangered—not just in the long run by a dying sun or some other cosmological cataclysm, but also in the short run. This view has emerged given the specter of human activity contributing to global climate change, proliferation of devastating weapons of mass destruction, all comingled with sectarian pathologies, and the seeming collective paralysis of governments to do anything about some of these threats. These realities have brought a certain collective synergy that has added to the gloom.

Never in the history of earth, until now, have humans possessed the means to end all life with weapons of mass destruction. Not only are nuclear weapons present in some of the most volatile regions of the world, but also the issue of proliferation is accelerating the magnitude of the problem. For these reasons a unique period has been entered, and it does not take a giant leap for the rational naturalistic mind to see the endgame. However, in the absence of meaning beyond the momentary existence it is an endgame that fits Alan Watts' description of life being little more than "a spark of light between one eternal darkness and another." This view can be reinforced by the daily headlines, sometimes suggesting that eternal darkness for all life on earth looms near indeed.

It is ironic that in the midst of the wonders of science and medicine, life now seems more tenuous than ever. Paradoxically, sectarian calls for dogma influenced order often seem to bring ever-greater disorder with dogmatic believers, eager to sacrifice innocent life in service to warped ideology.

From a scientific perspective life on planet earth in the long run ultimately does not survive. But there are now urgent questions about short-term biological survivability. It seems that from the available public evidence, the question is not "Will humans survive?" but only "How long?" The absurdity of a life of struggle, suffering and death is self-evident for beings born to love, hope and purpose. Yet this is the reality that public evidence presents, for while science has been a powerful catalyst to human insight and wisdom on matters related to how the universe functions, it has largely failed in providing assistance to existential meaning. If there is something other than the world available to sensate beings, it is not discoverable by empirical endeavors, and so those in search of purpose must therefore pursue other avenues.

By contrast, the tacit component inherent to revelation brings with it assumptions about the nature of reality that extends beyond the senses. In Christian theology, life is infused with a belief in God, one who is creator of the universal order, and one who offers eternal life. From such understandings a worldview is born that can provide context and meaning to life, much as a concerto achieves context and meaning, not from the isolated individual notes, but from a symphony of notes that come together through the efforts of a conductor. Yet transcendent matters can never be understood as part of legitimate scientific assumptions, given that they generally have no connection to empirical matters. In all cases they must be thought of as philosophical or theological in nature.

Even though the magisterium of sacred authority and that of nature largely run on distinct tracks, there are occasions when religious dogma wanders into areas that overlap with sense-based data. An example discussed in earlier chapters would include some interpretations of Genesis. First there will be a question as to whether these variant traditions are fundamentally at odds with each other, or whether there are ways they can be reconciled. If they are in fact irreconcilable, then either one or both are misinterpreting the reality—and possibly both. Certainly as a general rule they cannot both be viewed as fundamentally right. So it is this intersecting space that becomes a point of focus if there is a desire to optimize human understanding of reality, and in effect stay relevant and credible.

Many people coming from a faith tradition at variance with scientific understandings automatically assume that science cannot be right—right about earth and life on it having been created vastly longer than a few thousand years ago—even in the face of multiple streams of data that point clearly to those very conclusions. Yet because of the fundamental power of sense based endeavors and the corresponding success of the scientific enterprise, there would seem to be value for those who may be skeptical about this, to at least reflect strategically on the category of sacred assumptions and consider ways they can be respectful of sense based knowledge.

After all, within the framework of an epistemological triad of sense, reason, and revelation, some importance attaches to the maintenance of a grounded connection to the real world, particularly if it is understood that the tacit element must always be accountable to sense and reason wherever possible. In cases of overlapping interests within this triad, there is strategic importance for those connected to the sacred enterprise to remember that the goal is to achieve as close an approximation to reality as possible, and that therefore the sciences should be viewed as an ally towards this end.

All too often, in my experience, *sacred beliefs* are confused as *truth*—either by way of claim or practice—with the holder never recognizing that while *truth* represents the essence of the matter, *beliefs* can and do change as part of an organic human process of emergent understandings. Unless this very fundamental point is recognized, history has made clear that human nature tends to engage in programs that treat sacred beliefs in idolatrous ways, often becoming the tail that wags the dog. In such event, there might be an inappropriate inclination for dogma to ignore processes that could otherwise transform misunderstandings, turning ignorance into wisdom.

Conceptually, as an individual gets into the details of Christian narrative, it would, on the face of it, seem to have the capacity to elevate humans to an ordered existence, providing a pathway up out of the ontological swamp that is represented by the law of the jungle. In addition, it also transmits hope beyond death. Yet, as attractive as its conceptual framework may be in conveying meaning for vast numbers of people, history has demonstrated that on many occasions the Christian narrative has become consumed by a variety of tribal pathologies, resulting in some very petty dogma wars. While the larger narrative may be successful in conveying meaning, the history of religiously inspired warfare stands as proof that it does not always conveyed genuine benefits to humanity.

In the worst of cases, sectarian tribalism has been the catalyzing force in campaigns of genocidal death and destruction. Most readers will likely be unable to relate to the more extreme parts of this history, and yet go on to engage in other despicable, if less egregious, behavior in the name of God. There can be significant psychological power attached to a certainty of mind when acting as God's agent towards what most of the civilized world would view as a contemptible end. Much of this miserable history has reduced the world to tribal terms that go beyond *us* vs. *them*, to its more destructive manifestation of *good* vs. *evil*.

Unfortunately, the darker aspects of religious conviction have been evident throughout human history—and as Karen Armstrong has observed, it has been particularly a problem within monotheistic religions. While recent history may lead many to point an accusatory finger particularly at Islam, a more cerebral and sober analysis will reveal that the documented histories of both Judaism and Christianity prevent them from claiming a superior edge, even if outwardly more civilized in recent centuries. Even here, where the outward face seems more civil, just beneath the surface the pot simmers with strains of intolerance and fundamentalist extremism. Very often pious motives can act as the igniting force behind bigotry, inquisitions and/or acting out to force others (sometimes legislatively) to a particular point of view or behavior—sometimes while claiming to be engaged in doing God's work, acting as agents of His will for the world. As for Christianity itself, given its founding message of love and peace, it remains noteworthy the amount of spilled blood down through the ages of time, all done in the name of a loving God.

The irony of this history is reflected in the evidences of revelation itself, where ambiguity often reigns, and where competing ideas are sometimes based on murky provenance—things discussed earlier. Great horrors have been committed for ideas that at times were vague to begin with, and often not well grounded. The tough part of all this is that in many particulars the realm of the sacred extends beyond the reach of empirical verification, with only occasional points of intersection.

When issues do intersect, the choice of strategies range from a sacred interest pursuing common ground with the sciences, to one that represents a more arrogant and militant strain of religiosity that eschews science. When the latter happens, sacred interests are proceeding down a precarious trail, inclining to act in unaccountable ways by ignoring, or dismissing inconvenient data.

Interestingly, "certain minds" are essentially closed minds—closed to data; closed to the real world; closed to anyone holding a different view. History is quite clear on where such attitudes lead, often spinning out in egregious forms of sectarian strife. For sure, being "certain" provides a platform from which to judge everything else, but also unfortunately the context by which superstition becomes the guiding light, heightening the prospects that unmoored tribalism will descend down a destructive trail. In all of this there would seem to be a want of humility and wisdom within a tribal community so as to allow the sacred to connect with the real world where possible. As transparently inappropriate as it would seem to be, to approach the sacred with sanctimonious certainty, it is surprisingly common to find religious dogma packaged rather rigidly as "truth" even while perhaps asserting ideas that are ironically disconnected from real world data.

Thus, how the uncertainties of belief are resolved is terribly important to the human enterprise—particularly for beings that seek some overriding cosmic purpose and wish to do so with credibility. More than anything, these issues are an ever-present reminder of the need for modesty in how the storyline of mythos is embraced, for there should be great resistance to translate the evidences of faith into conceptual certitudes—particularly when doing so requires that sense-data be ignored, or treated in self-deceptive ways.

In order to keep the pursuit of truth in balance there is procedural significance attached to recognizing the tacit component as the foundation of all knowledge. But if the tacit component is to emerge in a credible way as to sacred concerns, then it must be held accountable to both rational and sensate knowing to the extent possible. This does not imply that any single part of the triad can be expected to be on the right track all the time, but is a simple acknowledgement that the sacred cannot be credible while ignoring other parts of the triad. The rational and sensate should therefore be expected to assume a role within a sacred narrative that addresses purpose. Such tools must remain as essential to the process of unraveling the nature of reality in a world that witnesses the ontological ubiquity and import that mythos transmits to the world.

On points of intersection when friction arises, the argument above is not one of capitulation by those holding sacred concerns—simply patience and humility. Time can work a lot of the frictions out.

In thinking about the plurality of all meaning-making paradigms, it is hard to miss the common characteristic of them all, that being that they essentially convey something of the human condition—that all mortals are thrashing about in the haze of reality, driven by the want of ontological context and understanding, with a desire for cosmic relevance in the grand scheme of things. For finite creatures with sentient and cognitive capacities, such issues come with some urgency, even though the most that can be expected is that reality will continue to only offer up ambiguities as a significant part of the equation. Humans simply have no choice but to live with a level of existential mystery and uncertainty if the realm of the transcendent is to be incorporated into a master narrative in any credible way. From this foundation may come certain private evidence—evidence that is personally meaningful, but which may not be universally recognizable, or accessible to systematic inquiry. As long as such appeals are modest, they may be of personal, if not universal value.

When all is said and done, there would appear to be great wisdom in understanding the limits of human knowledge regarding transcendent matters, thereby avoiding the seducing notion that *purpose* can only be found in rigid dogmatism. What is being proposed here is the possibility that from a more modest framework a sense of purpose can also take hold, and with it an ability to confront the sober realities of this world. While no one has all of the answers, if the tacit element includes a metaphysical component that can assist in providing direction to life it may inspire serenity and hope in the face of an uncertain world filled with adversity.

A Transcendent Hope

In spite of the darker aspects of religious dogmatism, sacred values continue to shape and mold culture, contributing to it many positive ideals, not the least of which would be the meaning that it can convey to life, as well as the role it plays in shaping the fabric of society. Yet, ambiguity, mystery, and uncertainty remain as a fundamental aspect of all sacred paradigms. Because of this, it is noteworthy that philosophers and theologians have created a rich series of logical formulations for putting the sacred narrative on an enhanced foundation. Some of these ideas were outlined in chapter 9, with a number of these constructs of seeming value. Yet ultimately none of them are capable of moving knowledge of God beyond the level of faith, with the sources coming from that which is revealed in sacred writings, the observations of nature (assumed to be the handiwork of God), logic and sensate considerations, and finally, personal encounters that may be interpreted as having a connection to the divine.

No one can explain with complete satisfaction the mysteries of human existential reality, or of the sensate hiddenness of God, or how a loving God can coexist with a broken world of suffering, or of the mystery of moral good and evil. Apart from sacred texts, this latter category can often be evaluated by the instinct for survival and order. In this indirect way, the capacity to evaluate some expressions of good and evil can be made in terms of harmful effects. But in the absence of adequate empirical tools on all these issues, they must be recognized as among the remaining loose ends (parenthetically the tacit element creates loose-ends for all knowledge). Yet in spite of this unfinished business many have found enough viability in the Christian master narrative so as to live with hope and faith that reality at a fundamental level is consistent with the essence of this narrative. Certainly, the central Christian theme of love and respect towards others would seem to be a prerequisite for a world free from suffering, and to the extent that these values are practiced in the here and now, they can only add commensurately to the well-being of all sensate creatures.

The narrative that has inspired hope for billions of people through the millennia of time lives on to capture the imagination and passions of many more. It has been portrayed very effectively through the arts, through grand architecture, and through compelling music. Michelangelo, Raphael, Leonardo da Vinci, Johann Sebastian Bach, George Fredric Handel and a great many others have been an important part of a rich tradition that has elevated the human spirit up out of an ontological swamp. The universal message of these great masters has been to convey the language of a transcendent hope.

This transcendent hope has the capacity to be something more than an empty fantasy, as it can be supported by the realities of nature that are pregnant with possibilities. Among these are the complex coding that is the basis of all life, as well as the seeming mathematical odds involved of human reality being merely an accident of random processes outside of a larger framework of understanding. The prospect of facing the data by reflecting on how functional complexity could have evolve from a primordial singularity involving inorganic chemical compounds, seems dauntingly pointless given the apparent improbabilities. High on many lists will therefore be a tacit bias of some form of strategic, purposeful design.

It was the Apostle Paul who spoke of "faith, hope and love" as part of a package, and if engaged strategically, it is arguable that life in the aggregate would not be improved upon by living apart from this—particularly in view of the difficulty humans have in finding deep ontological purpose aside from that which can be found in the sacred.[3] In addition, the act of internalizing these elements may arguably become a side benefit of enhanced well-being.

One of the more significant points to this discussion has been to find a way in which to connect credibility to "faith," "hope," and "love," and as has been proposed this can be accomplished by remaining mindful of the tacit foundation upon which the narrative must be built. A program of credibility will necessarily require the exercise of restraint in issuing truth-claims that may be at variance with the data. *Hope* can desire that an increasingly sophisticated understanding of data that presently vexes a faith tradition may eventually mature in ways that may be more consistent with that tradition, but hope cannot *will* such an outcome. Most importantly, any religious community that is committed to the notion of "doctrinal truth" will have no choice but to proceed in ways that maintains respect for all sensate insights. This can be accomplished only by patience.

It has been my goal in the foregoing discussion to elevate a more modest, defensible, and strategic approach to sacred ideas. I have attempted to lay out in a reasonably systematic way the issues that arise from certitudes in the literal reading and interpretation of all sacred texts, including the Genesis creation account. In setting dogmatism and mental certitudes aside, it seems that the most important goal must be the correspondence of belief to the reality. In earlier centuries the teaching magisterium of the Church could set forth dogma to its members, who for the most part were simple people lacking any ability to mount credible challenges. It is a vastly different world today. With the exception of those who imbibe at the fountain of postmodern superstition, this is an age of free flowing and accessible information, and a religious magisterium of any faith community that attempts to set forth dogma that runs against the dominant interpretations of data will find the reality more difficult to subvert than it would have been in an earlier age.

In the contemporary world the senses are considered to be a credible discerner of empirical realities, even if it cannot guarantee certainty. Such inquiries can evaluate data by logical abductions, which is a method of tracing processes of the natural order back in time to explain a present manifestation. It is in this way, for example, the big bang cosmology employs such tools by simply putting the observed expanding universe in reverse motion and concluding that energy and matter all came together to form a singularity at a remote primordial point in time. Obviously, there is no way to ever prove this past history beyond all doubt, but at this stage it

---

3. See generally, 1 Cor 13.

does appear to be a reasonable supposition—a supposition of beginnings that is not inconsistent with a strategic interpretation of the Genesis account of beginnings. Likewise, as science attempts to piece together the formation of earth and the solar system, it is possible to reach certain conclusions based on observations of the birth of new solar systems in distant nebula. The conclusions made from such observations do not so much overturn biblical insights as they do to add spectacular new details to that ancient narrative. It is now clear that creation was not just a one-time event, but is an ongoing dynamic process.

Over against the guardians of orthodoxy who refused to allow the light of the sensate to shine in on sacred writings—light that has the capacity to render up profound if incomplete insights—there are empirical tools that can be brought to bear that may actually enhance human understanding of revelation. But unless a strategic vision is captured in a way that can pump new life into a flaccid orthodoxy, narrow-mindedness may exact a steep price as the sobering march of history and knowledge become the final arbiter—that is, if relevance and credibility should matter at all.

The path forward that has been proposed over the intervening pages has been to commit to the processes by which knowledge is acquired; process that can transform what might otherwise be labeled *superstition* and convert it into a credible, systematic framework of understanding. It is the commitment to such a process that can deliver doctrinal integrity and credibility to metaphysical concerns by treating "sacred truth" as dynamic rather than as static.

It was Albert Einstein who said, "The most beautiful thing we can experience is the mysterious. It is the source of all true art and all science." If this statement could be improved upon, it would be to include within it, the mystery of metaphysical contemplations—that the most beautiful thing we can experience is the mystery associated with the sacred and transcendent. Such mystery adds to the intrigue of life, with some of the connecting details serving as waymarks for those who choose to live with hope and faith, in the face of uncertainty.

Appendix A

In Ellen White's earliest account of earth's history, she attributed her views to a vision that she and her followers believed to be from God. In a recently published biography on Ellen White, titled, *Ellen Harmon White: American Prophet*, notes the following:

> Ellen White claimed that God had taken her back in vision to witness the actual creation of the world. Pleased that the Lord had made her "his humble instrument in shedding some rays of precious light upon the past," she eagerly grasped the opportunity to flesh out the brief history given in the Bible. During her out-of-body experience she had seen "that the first week, in which God performed the work of creation in six days and rested on the seventh day, was just like every other week." She also confirmed "that the world is now only about six thousand years old."[1]

The authors note that this may have been the first of about 18 times that she referred to the earth's age of about 6000 years.

In Ellen White's own words is the following statement:

> "Infidel geologists claim that the world is very much older than the Bible record makes it. They reject the Bible record, because of those things which are to them evidences from the earth itself, that the world has existed tens of thousands of years. And many who profess to believe the Bible record are at a loss to account for wonderful things which are found in the earth, with the view that creation week was only seven literal days, and that the world is now only about six thousand years old. These, to free themselves of difficulties thrown in their way by infidel geologists, adopt the view that the six days of creation were six

---

1. Numbers, Ronald L., and Rennie B. Schoepflin, Ed Terrie Dopp Aamodt, Gary Land, and Ronald L Numbers, *Ellen Harmon White: American Prophet*, Oxford University Press (2014), 214. The source document for this quotation is cited as: White, *Spiritual Gifts* (1864), 90-92.

vast, indefinite periods, and the day of God's rest was another indefinite period; making senseless the fourth commandment of God's holy law. Some eagerly receive this position, for it destroys the force of the fourth commandment, and they feel a freedom from its claims upon them . . . ."

"I have been shown that without Bible history, geology can prove nothing. Relics found in the earth do give evidence of a state of things differing in many respects from the present. But the time of their existence, and how long a period these things have been in the earth, are only to be understood by Bible history. It may be innocent to conjecture beyond Bible history, if our suppositions do not contradict the facts found in the sacred Scriptures. But when men leave the word of God in regard to the history of creation, and seek to account for God's creative works upon natural principles, they are upon a boundless ocean of uncertainty. Just how God accomplished the work of creation in six literal days he has never revealed to mortals. His creative works are just as incomprehensible as his existence."[2]

While, it is quite likely that some of the early geologist were agnostic on religious matters, the above statement is woefully inadequate in as much as it appears to presuppose that geologists as a class were all infidels. The reality is, as developed by Davis A. Young and Ralph F. Stearley in the extensively footnoted, *The Bible, Rocks and Time*, a great many of the leading geologists of the sixteenth through eighteenth centuries were Christians who began with the presupposition of the age of the earth being about 6,000 years old—generally traced to Ussher's Chronology of the biblical patriarchs.

As evidence accrued, these early geologists were often troubled by the message that seemed to be conveyed by the geological evidence of a far older earth. They often struggled with reconciling common Christian dogma with geological evidence. But as consistent evidence continued to accrue for an old earth, it became necessary to eventually abandon the notions of a young earth.

The point, here, is that, broadly painting geologists who had concluded that the earth was far older than 6,000 years, as being "infidels" is simply a harsh misrepresentation of the facts—and to be charitable—perhaps innocently so.

From the perspective of the 21st century it is also inaccurate to conclude that geologists can prove nothing without the bible, for the reality is that geological evidence has the capacity to speak for itself on many levels.

---

2. White, Ellen G. *Spiritual Gifts* Vol. 3, 90 – 93

Appendix B

The original wording of Fundamental Belief #6: "God is Creator of all things, and has revealed in Scripture the authentic account of His creative activity. In six days the Lord made "the heaven and the earth" and all living things upon the earth, and rested on the seventh day of that first week. Thus He established the Sabbath as a perpetual memorial of His completed creative work. The first man and woman were made in the image of God as the crowning work of Creation, given dominion over the world, and charged with responsibility to care for it. When the world was finished it was "very good," declaring the glory of God." (Gen. 1; 2; Ex. 20:8-11; Ps. 19:1-6; 33:6, 9; 104; Heb. 11:3.); **The revised wording of Fundamental Belief #6** (emphasis added): "God is the Creator of all things. He has revealed in Scripture the authentic and historical account of His creative activity. In a **recent** six-day creation the Lord made "the heavens and the earth, the sea, and all that is in them" and rested on the seventh day. Thus He established the Sabbath as a perpetual memorial of His creative work, **performed and completed during six literal days that together with the Sabbath constituted the same unit of time that we call a week today.** The first man and woman were made in the image of God as the crowning work of Creation, given dominion over the world, and charged with responsibility to care for it. When the world was finished it was "very good," declaring the glory of God. (Gen. 1-2; 5; 11; Ex. 20:1 8-11; Ps. 19:1-6; 33:6, 9; 104; 2 Isa. 45:12; Acts 17:24; Col. 1:16; Heb. 11:3; Rev. 10:6; 14:7.)"

Bibliography

Achenbach, Joel. "The Age of Disbelief," *National Geographic,* March 2015, 45.
Adams, Fred, and Gregory Laughlin. *The Five Stages of the Universe: Inside the Physics of Eternity.* New York: Simon & Schuster, 1999.
Adler, Mortimer. *Ten Philosophical Mistakes.* New Jersey: Macmillan, 1987.
Armstrong, Karen. *The Bible.* Grove Press: New York, 2007; see also *The Battle for God: A History of Fundamentalism*, Alfred A. Knoft, 2000; see also, *The History of God*, New York: Alfred A. Knopf, 1993.
Augustine. *Confessions.* Trans. H. Chadwick, New York: Oxford University Press, 1991.
Barbour, Ian G. *Religion and Science: Historical and Contemporary Issues.* New York: HarperCollins, 1997.
Barker, S. A., J. A. & Christian, S. T. N. N-di-methyl-tryptamine: an endogenous hallucinogen. *Int Rev Neurobiol*, 1981.
Bernstein, Jay H. *The Data-Information-Knowledge-Wisdom Hierarchy and its Antithesis* http://journals.lib.washington.edu/index.php/nasko/article/viewFile/12806/11288, 69 - 70
Boyd, Gregory A. *God at War.* Downers Grove, IL: InterVarsity, 1997.
Brown, James Robert. *Who Rules in Science.* Cambridge, MA: Harvard University Press, 2001.
Cirignotta, F., Tudesco, C. V. P & Lugaresei, E. Temporal lobe epilepsy with ecstatic seizures (so called Dostoevsky's epilepsy). *Epilepsia*, 1980.
Clayton, Philip & John Knapp. *The Predicament of Belief.* Oxford University Press, 2011.
Comings, David. *Did Man Create God?* Duarte, CA: Hope, 2008.
Davidson, Richard M. "The Authority of Scripture: A Personal Pilgrimage." *In Journal of Adventist Theological Society* 1/1 (1990) 39–56.
Eakin, Emily. *New York Times.* "So God's Really is in the Details?" May 11, 2002. http://www.nytimes.com/2002/05/11/arts/so-god-s-really-in-the-details.html?pagewanted=all&src=pm.
Enns, Peter. *Inspiration and Incarnation*, 2d Ed., Grand Rapids, MI: Baker Academic, 2015
Erhman, Bart. *God's Problem*, New York: HarperCollins, 2008;
———. *Misquoting Jesus*, New York: HarperCollins, 2005.
Falk, Dan. *In Search of Time*, New York: Thomas Dunn Books, St. Martin (2008). See specifically chapter titled, "All Things Must Pass," 245–271.
Gilovich, Thomas. *How We Know What Isn't So,* New York: The Free Press (a division of Simon & Schuster), 1991.

Geisler, Norman. *Baker Encyclopedia of Christian Apologetics,* Grand Rapids, MI: Baker Books, 1999.

Gould, Stephen J. and Jonathan Cape, *Leonardo's Mountain of Clams and the Diet of Worms: Essays on Natural History,* Cambridge, MA: First Harvard University Press Ed., 2011.

———. *Rocks of Ages: Science and Religion in the Fullness of Life.* New York: Ballantine Books, 2002.

Hand, David J. *The Improbability Principle: Why Coincidences, Miracles, and Rare Events Happen Every Day,* New York: Scientific American/ Farrar, Straus and Giroux, 2014.

Havener, Timothy. God on Trial: The Verdict, https://www.youtube.com/watch?v =dx7irFN2gdI

Heidegger, *Being and Time,* New York: HarperCollins, 1962.

Hicks, John. *Evil and the God of Love* San Francisco, CA: Harper & Row, 1978.

Holt, Jim. *Why Does the World Exist?* New York: Liveright, a division of W. W. Norton & Company, 2012.

Kauffman, Stuart. *Reinventing the Sacred.* New York: Basic Books, Perseus Book Group, 2008.

Kootsey, Mailen. What makes Science Valuable, Spectrum Magazine, http://spectrummagazine.org/blog/2012/07/09/bringing-real-world-genesis-what-makes-science-valuable-part-1-observations-experime, 2012 See also http://atoday.org/?s=ultra+darwinists+and+conservative+christians+agree

Kreeft, Peter. The Argument from Pascal's Wager, http://peterkreeft.com/topics/pascals-wager.htm

Mackey, Charles, *Extraordinary Popular Delusions and the Madness of Crowds,* Berlin: Renaissance Classics, 2012; first published in 1841.

Meyer, Stephen. *Signature in the Cell,* New York: HarperCollins, 2009.

Mlodinow, Leonard. *The Drunkard's Walk: How Randomness Rules our Lives,* New York: Pantheon Books, 2008.

Numbers, Ronald L., and Rennie B. Schoepflin, Ed Terrie Dopp Aamodt, Gary Land, and Ronald L Numbers, *Ellen Harmon White: American Prophet,* Oxford: Oxford University Press, 2014.

Paley, William. *Natural Theology,* chs.1–3

Patterson, McMillan, Switzler, et al. *Crucial Conversations,* Columbus, OH: McGraw-Hill Education, 2011

Penfield, Wilder. The Role of the Temporal Cortex in certain Psychical Phenomena. *The Journal of Mental Science* (1955) 424.

———. Penfield, W. & Perot. The Brains' Record of Auditory and Visual Experience, *Brain,* (1963) 696.

———. The Permanent Record of the Stream of Consciousness. Acta Psychol (Amst). (1955) 11:47-69.

Pensinger, M.A. Striking EEG profiles from single episodes of glossolalia and transcendental meditation. *Percept Mot Skills.* (1984) 58: 127–233.

———. Neuropsychological Basis of God Beliefs. Prager, New York, 1987.

Pinnock, Clark H., Rice, Richard, John Sanders, William Hasker and David Basinger. *The Openness of God: A Biblical Challenge to the Traditional Understanding of God,* 1997.

Planck, Max. *Where Is Science Going?* Trans. J.S Murphy, London, Allen and Unwin, 1933.
Plantinga, Alvin. *Warranted Christian Belief,* New York: Oxford University Press, 2000.
Polanyi, Michael, *Personal Knowledge,* University of Chicago Press, 1962.
Punkt, Sammel. "Probability Theories and the Justification of Theism," http://sammelpunkt.philo.at:8080/1679/1/portugal.pdf
Rees, Martin. *Our Final Hour,* New York: Basic Books, Perseus Book Group, 2003.
Riojas, Gabriel M. "A House of Idols," *Spectrum Magazine*, www.spectrummagazine.org, October 1, 2013.
Rovelli, Carlo. *The First Scientist: Anaximander and His Legacy.* Yardley, PA: Westholme, 2011.
Rowley, Jennifer and Richard Hartley, Organizing Knowledge" *An Introduction to Managing Access to Information.* Burlington, VT: Ashgate Publishing Limited, 2006.
Sacks, Jonathan. *The Great Partnership: Science, Religion, and the Search for Meaning,* New York: Schocken Books, 2011.
Sadeghian, Abbas. *Sword and Seizure: Muhammad's Epilepsy & Creation* of Islam, Enumclaw, WA: Annotation, 2006.
Saletan, William. "God's Work?" *Slate* http://www.slate.com/articles/health_and_science/human_nature/2014/12/creationism_poll_how_many_americans_believe_the_bible_is_literal_inerrant.html
Schoeder, Gerald L. *God According to God,* New York: HarperCollins, 2009.
Seat, Leroy, Robert E Patterson ed., *Science Faith and Revelation*, Nashville, TN: Broadman, 336-354, 1979.
Slick, Matt. The Cosmological Argument, http://carm.org/cosmological-argument;
Stark, Thom. *The Human Faces of God,* Eugene, Oregon: Wipf & Stock, 2011.
Taylor, Charles. *A Secular Age*, Stanford University Press, 2007.
Templeton, Charles. *Farewell to God,* McClelland & Stewart, Toronto, Canada, 1996.
Thompson, Damian, *Counterknowledge.* New York: W.W. Norton, 2008.
Ussher's Chronology, http://en.wikipedia.org/wiki/Ussher_chronology
Wallace, Danny, "Knowledge Management: Historical and Cross-Disciplinary Themes," Westport, CT: *Libraries Unlimited, 2007.*
Watt, Alan W. *The Wisdom of Insecurity*, Pantheon Books, 1951.
Welles, James F. *Understanding Stupidity*, Mount Pleasant, 1986.
White, Ellen G. *Spiritual Gifts,* 1864.
Wilbur, David W. *Power and Illusion,* 2010.
Wilson, Edward O. *Consilience: The Unity of Knowledge*, Alfred A. Knopf: New York, 1998.
Yi, Zane G. *The Possibility of God: An Examination and Evaluation of Charles Taylor's Transcendental Critique of Closed Worlds*, A Dissertation, Fordham University
Yockey, Hubert P. *Information Theory, Evolution, and the Origin of Life* (Cambridge University Press) 2005.
Young, Davis A. and Ralph F. Stearley. *The Bible, Rocks and Time.* Downers Grove, IL: IVP Academic, 2008.
Zin, Chain, "Conceptual Approaches to Defining Data, Information, and Knowledge," *Journal of the American Society for Information Science and Technology* 58 (4): 479–493, 2007.

Subject Index

Adventist, xiii–xv, xvii, xx–xxi, 64, 183
Agnostic, xxxii, 117, 119, 125, 129, 138, 152, 180
Agnosticism, 21, 69, 81, 125, 138, 152
Apostle Paul, 87, 93, 175
Armstrong, Karen, xviii, xxix–xxxn16, 137n41, 140, 152–153n9, 162, 172

Bayes Theorem, 127, 139
Bayesian, 127–128
Bernstein, Jay, 21–22, 183
Boyd, Gregory, 83n7, 84–85

Catholic, xiii
Clayton, Philip, 78, 93, 183
Comings, David, 95–103, 183
Copernican, xxix–xxxn16, 40, 109
Copernicus, x, 38, 72, 109, 167
Coral Reef, 34
Counter-knowledge, 20–21, 24–25

Demski, William, 131, 136–137
Determinism, 55–56, 131
DIKW Pyramid, 10–11n4, 21–22
Divine Sovereignty, 49, 52, 55-56, 83–84n10
Divine Will, 52, 54–55, 90, 92
DNA, xix, 45–46, 132–137n40, 140, 146

Ehrman, Bart, 68–72, 83n6, 143n5
Explicit Knowledge, 109, 155, 162

Fundamentalism, xv, xviii–xix, xxix–xxxi, 116, 142, 183

Galileo, x, xxxi, 40, 50, 109, 145, 158, 167
Genesis, xvi, xix, xxi–xxii, xxixn14, xxxiii, 23, 26, 30–32, 46n5, 74, 87, 93, 114, 128, 146, 151, 158–159, 171, 176–177, 184
Geologic column, xx, 34
Geoscience Research Institute, xvii, xix
Golden Rule, 141, 152–153
Gould, Stephen Jay, 135n35, 148–149, 184
Greek, 50–52, 71–72, 83, 90, 93, 112

Hand, David J., 130, 132n27, 184
Hebrew, 51, 69, 90, 93, 112
Hicks, John, 84, 184
Hermeneutics, 59n2, 64–65
Hypothesis, xvii, 22, 25, 36–40, 43, 101, 109, 117, 127–128, 135–136, 139

Ice cores, xx, 33
Idolatry, v, xxxii, 58, 61, 72, 74
Indeterminacy, 8, 55–56
Inerrancy, xv, 58, 64, 68–70, 72, 75, 143–144

Jew, xiii,
Jewish, 51, 69, 112, 129
Judaism, xxvii, 13, 63, 80, 104, 172

Knapp, Steven, 78, 93, 183
Knowledge, v, viii–x, xv, xxii, xxviii,, xxxi, xxxiv, 3–14
Kuhn, Thomas, 35

Light speed, xx, 32
Livio, Mario, 50

Mackay, Charles, 22
Magisterium, 145, 147–149, 171, 176
Magisteria, 148–151
Master Narrative, x, 4, 15, 24, 26, 107–108, 111, 113–116, 118–119, 124, 137, 139, 142, 146, 149–152, 154, 157, 161, 165, 167–168, 174–175
Mathematical, 37–38, 46, 49–52, 54–56, 78, 90–92, 123, 127, 130, 133, 137, 139, 175
Mesopotamian, 50–51, 90, 93
Meyer, Stephen, 131–133, 135–137, 184
Milton, John, 85
Mlodinow, Leonard, 51–52, 90, 184
Mormon, xiii
Muslim, xiii, 62–64, 112–113

Natural theology, 119, 126, 129n20, 137, 139, 184
Naturalism
  Methodological, 46 108, 114, 120, 149, 152
  Philosophical, 46, 152
NOMA, 148–150
Non-knowledge, 22

Occam's Razor, 46
Ocean Sedimentation, xx, 34
Ontological, xxvii, 14, 122–124, 168, 172–175
Opinion, x, xvi, xx, 3, 6–10, 14, 16, 79, 109, 119, 128, 139, 144, 160, 166

Pagel, Elaine, 86
Paley, William, 128, 129n20, 133, 184
Paradise Lost, 85
Pascal, Blaise, 125–126

Pascal's wager, 125–126, 184
Plantinga, Alvin, 123–124, 139, 151, 185
Polanyi, Michael, x, 108–110, 112, 114, 116–118, 128n19, 141, 185
Pragmatism, 6, 108, 119, 125
Presupposition, xi, xv, 12–13,18, 26–27, 30, 60, 69, 108–110, 117, 122, 124, 139–142, 151, 158–159, 168, 180
Probability, 8, 52, 54–57, 90–92, 125, 127–128, 130–131, 133, 136–137, 139, 184–185
Ptolemaic, x, 109, 167
Ptolemy, 109

Qur'an, 10, 62, 113

Radiometric Dating, xx, 32, 35
Reductionism, 45
Religious Fidelity, 119, 121–122

Saint Augustine, 83n7, 52, 183
Saint Thomas Aquinas, 128
Sagan, Carl, xxii
Satan, 86
Schroeder, Gerald L., 129–130
Seismic data, 34
Shannon, Claude, 131–134
Sola Scriptura, xvi, xix, 10n2, 58, 59, 63, 147
Swinburne, Richard, 126–128, 131

Tacit Knowledge/Element/Component, x, xi, 16–19, 23, 25, 78, 107, 109, 113–118, 128, 137, 139, 141, 148–149, 151, 155, 158, 162, 164–165, 168, 171, 173–176
Taylor, Charles, 110, 112, 114–115, 118, 185
Templeton, Charles, 81–82, 84, 185
Temporal Lobe Epilepsy, xx, 98–99, 183
Theological Rationalism, 119, 122
Theology of Causation, 119, 124, 139

Theory, 10, 38–40, 43–44, 47, 51–52, 90, 108–110, 119n1, 125, 129, 131, 133–134, 150, 185
Thompson, Damian, 20, 23, 185
TLE, xx, 98–99, 101–104
Torah, xiii, 51, 65, 145
Tribal, v, 16, 153–157, 161–163, 168–169, 172–173

Tribalism, xxvii, 155, 159, 162, 172–173

Wegener, Alfred, 34
White, Ellen, xiii–xvii, xix–xxi, 84, 179–180, 185

Yockey, Hubert, 134, 137, 185